I0021028

Simplifying Android Development with Coroutines and Flows

Learn how to use Kotlin coroutines and the flow API to handle data streams asynchronously in your Android app

Jomar Tigcal

BIRMINGHAM—MUMBAI

Simplifying Android Development with Coroutines and Flows

Copyright © 2022 Packt Publishing

All rights reserved. No part of this book may be reproduced, stored in a retrieval system, or transmitted in any form or by any means, without the prior written permission of the publisher, except in the case of brief quotations embedded in critical articles or reviews.

Every effort has been made in the preparation of this book to ensure the accuracy of the information presented. However, the information contained in this book is sold without warranty, either express or implied. Neither the author, nor Packt Publishing or its dealers and distributors, will be held liable for any damages caused or alleged to have been caused directly or indirectly by this book.

Packt Publishing has endeavored to provide trademark information about all of the companies and products mentioned in this book by the appropriate use of capitals. However, Packt Publishing cannot guarantee the accuracy of this information.

Group Product Manager: Rohit Rajkumar
Publishing Product Manager: Nitin Nainani
Senior Editor: Aamir Ahmed
Content Development Editor: Rakhi Patel
Technical Editor: Joseph Aloocaran
Copy Editor: Safis Editing
Project Coordinator: Manthan Patel
Proofreader: Safis Editing
Indexer: Manju Arasan
Production Designer: Sinhayna Bais
Marketing Coordinator: Teny Thomas

First published: July 2022

Production reference: 1220722

Published by Packt Publishing Ltd.
Livery Place
35 Livery Street
Birmingham
B3 2PB, UK.

ISBN 978-1-80181-624-3

www.packt.com

To my loving wife Celine for her support and encouragement, especially during the time I was writing this book.

To my parents for all their sacrifices and for raising me well.

– Jomar Tigcal

Foreword

I have known Jomar for over a decade. He and I met when I was planting the seeds for the Google Developer Groups in the Philippines. It started as a help group for Filipino developers who wanted to learn how to develop apps using Google technology. It was soon called the Google Technology Users Group, and it was eventually renamed as **Google Developer Group (GDG)**.

Jomar played a key role in the growth and development of this group. I was looking for someone who could help teach the community how to use HTML5. Jomar was then very shy, but I saw the spark in his eyes whenever new technology was introduced. After much convincing, Jomar agreed to conduct the training session.

We were really surprised by the number of attendees who joined the session. We even had more attendees who joined online (streamed with spotty internet through my tiny netbook). Jomar was obviously very nervous, but when he started getting into tech, I saw him blossom into the well-admired developer he is today.. Soon enough, he was conducting similar sessions during GDG events in Mountain View.

Jomar, with the rest of the GDG team, carried the torch for many years. He developed several useful applications like Sweldong Pinoy, Budget Pinoy, Thirteenth Month, Boto Ko, PHostpaid, Pinoy Jokes, and GDG apps (GDG Watchface and GDG Philippines). He continued to teach and share his knowledge with developers across the country. This book, where you will learn how to build high-quality and maintainable Android applications using Coroutines and Flows, did not come as a surprise.

Jomar continues to inspire and demonstrate to other developers that anything is possible if you put your mind to it. I'm no longer guilty of forcing Jomar to host that first HTML5 training. It is his destiny to help, encourage, and inspire other developers to learn and create more useful applications.

Aileen Apolo-de Jesus

Former Senior Program Manager, Google

Contributors

About the author

Jomar Tigcal is an Android developer with over 10 years of experience in mobile and software development. He has worked on various stages of app development for both small startups and large companies. Jomar has also given talks and conducted training and workshops on Android. In his free time, he likes running and reading. He lives in Vancouver, BC, Canada with his wife, Celine.

About the reviewer

Shreyas Patil is a Google Developers Expert for Android and works as an Android Developer at Paytm Insider. He is a self-taught developer and has experience in developing Android and web frontend and backend applications. He has several of his own applications published on the Google Play Store. He loves building products and apps and developing useful libraries that help everyone. He's also an organizer of the community of Kotlin Mumbai. He spends a lot of his development time contributing to open source projects, most of which he started by himself and can be found on his GitHub profile. He loves contributing to the community by writing blogs and articles and making open source projects that can help fellow developers to learn new things.

Table of Contents

3

Handling Coroutine Cancelations and Exceptions

4

Testing Kotlin Coroutines

Part 2 – Kotlin Flows on Android

5

Using Kotlin Flows

6

Handling Flow Cancelations and Exceptions

7

Testing Kotlin Flows

Index

Other Books You May Enjoy

Preface

Kotlin coroutines and flows allow developers to do asynchronous programming in Android using simple, modern, and testable code.

This book focuses on coroutines and flows using hands-on learning. You will begin with the basics of asynchronous programming, including an overview of coroutines and flows while also integrating them into your Android projects. You'll understand how to manage cancelations and exceptions, and then explore how to test your coroutines and flows.

By the end of this book, you will be able to use Kotlin coroutines and flows to simplify asynchronous programming in Android.

Who this book is for

This book is for Android developers who want to build high-quality apps using coroutines and flows and level up their Android development skills. Beginners with basic knowledge of Android development and Kotlin will also find this book useful.

What this book covers

Chapter 1, *Introduction to Asynchronous Programming in Android*, visits asynchronous programming in Android and shows the various ways it is being done now. Toward the end, the new recommended ways of coroutines and flows will be introduced.

Chapter 2, *Understanding Kotlin Coroutines*, introduces Kotlin coroutines and shows how they can be used for asynchronous programming in Android. It will demonstrate how to create coroutines, and discuss coroutine builders, scopes, dispatchers, contexts, and jobs.

Chapter 3, *Handling Coroutine Cancelations and Exceptions*, discusses coroutine cancelations, and how you can manage coroutine cancelations, timeouts, and exceptions properly.

Chapter 4, *Testing Kotlin Coroutines*, describes how you can test Kotlin coroutines in Android.

Chapter 5, *Using Kotlin Flows*, covers the basics of Kotlin flows and how they can be used for asynchronous programming in Android. It continues with creating flows with flow builders. It will also discuss flow operators, buffering and combining flows, and StateFlow and SharedFlow.

Chapter 6, Handling Flow Cancelations and Exceptions, explores how to manage cancelations, completion, and exceptions in your flows.

Chapter 7, Testing Kotlin Flows, provides details on how you can test the flows in your Android project.

To get the most out of this book

You will need to have basic Android development skills and knowledge of using Kotlin.

You will need to have the latest version of Android Studio. You can download the latest version at `https://developer.android.com/studio`. For an optimal experience, the following specifications are recommended:

- Intel Core i5 or equivalent or higher
- 4 GB RAM minimum
- 4 GB available space

Software/hardware covered in the book	Operating system requirements
Android Studio	Windows, macOS, or Linux

If you are using the digital version of this book, we advise you to type the code yourself or access the code from the book's GitHub repository (a link is available in the next section). Doing so will help you avoid any potential errors related to the copying and pasting of code.

Download the example code files

You can download the example code files for this book from GitHub at `https://github.com/PacktPublishing/Simplifying-Android-Development-with-Coroutines-and-Flows/`. If there's an update to the code, it will be updated in the GitHub repository.

We also have other code bundles from our rich catalog of books and videos available at `https://github.com/PacktPublishing/`. Check them out!

Conventions used

There are a number of text conventions used throughout this book.

`Code in text`: Indicates code words in text, database table names, folder names, filenames, file extensions, pathnames, dummy URLs, user input, and Twitter handles. Here is an example: "`runOnUIThread` will perform the `displayText(text)` function on the main UI thread."

A block of code is set as follows:

```
lifecycleScope.launch(Dispatchers.IO) {
    val fetchedText = fetchText()

    withContext(Dispatchers.Main) {
        displayText(fetchedText)
    }
}
```

When we wish to draw your attention to a particular part of a code block, the relevant lines or items are set in bold:

```
private fun fetchTextWithThread() {
    Thread {
        // get text from network
        val text = getTextFromNetwork()

        runOnUiThread {
            // Display on UI
            displayText(text)
        }
    }.start()
}
```

Any command-line input or output is written as follows:

```
java.lang.IllegalStateException: Module with the Main
dispatcher had failed to initialize. For tests Dispatchers.
setMain from kotlinx-coroutines-test module can be used
```

Bold: Indicates a new term, an important word, or words that you see onscreen. For instance, words in menus or dialog boxes appear in **bold**. Here is an example: "In Android Studio, the **Editor** window identifies the suspending function calls in your code with a gutter icon next to the line number."

> **Tips or Important Notes**
> Appear like this.

Get in touch

Feedback from our readers is always welcome.

General feedback: If you have questions about any aspect of this book, email us at `customercare@packtpub.com` and mention the book title in the subject of your message.

Errata: Although we have taken every care to ensure the accuracy of our content, mistakes do happen. If you have found a mistake in this book, we would be grateful if you would report this to us. Please visit `www.packtpub.com/support/errata` and fill in the form.

Piracy: If you come across any illegal copies of our works in any form on the internet, we would be grateful if you would provide us with the location address or website name. Please contact us at `copyright@packt.com` with a link to the material.

If you are interested in becoming an author: If there is a topic that you have expertise in and you are interested in either writing or contributing to a book, please visit `authors.packtpub.com`

Share Your Thoughts

Once you've read *Simplifying Android Development with Coroutines and Flows*, we'd love to hear your thoughts! Scan the QR code below to go straight to the Amazon review page for this book and share your feedback.

`https://packt.link/r/1801816247`

Your review is important to us and the tech community and will help us make sure we're delivering excellent quality content..

Part 1 –
Kotlin Coroutines
on Android

In this part, we will introduce the concept of asynchronous programming and discuss the new and recommended way to do it with Coroutines. We'll learn how to create coroutines, handle cancelations and exceptions, and test them.

This section comprises the following chapters:

- *Chapter 1, Introduction to Asynchronous Programming in Android*
- *Chapter 2, Understanding Kotlin Coroutines*
- *Chapter 3, Handling Coroutine Cancelations and Exceptions*
- *Chapter 4, Testing Kotlin Coroutines*

1
Introduction to Asynchronous Programming in Android

There are Android applications that work on their own. But most apps retrieve data from or send data to a local database or a backend server. Examples of these include fetching posts from a social network, saving your favorites from a list, uploading an image, or updating your profile information. These tasks and other resource-intensive computations may happen instantly or take a while to finish. Factors such as internet connection, device specifications, and server settings affect how long these operations take.

Long-running operations must not be performed on the main UI thread as the application will be blocked until they are completed. The application might become unresponsive to the users. Users may not be aware of what's happening, and this might prompt them to close the app and reopen it (canceling the original task or doing it again). The app can also suddenly crash. Some users might even stop using your app if this happens frequently.

To prevent this from happening, you need to use asynchronous programming. Tasks that can take an indefinite amount of time must be done asynchronously. They must run in the background, parallel to other tasks. For example, while posting information to your backend server, the app displays the UI, which the users can interact with. When the operation finishes, you can then update the UI or notify the users (with a dialog or a snackbar message).

With this book, you will learn how to simplify asynchronous programming in Android using Kotlin coroutines and flows.

In this chapter, you will first start by revisiting the concept of asynchronous programming. After that, you will look into the various ways it is being done now in Android and how they may no longer be the best way moving forward. Then, you will be introduced to the new, recommended way of performing asynchronous programming in Android: coroutines and flows.

This chapter covers three main topics:

- Asynchronous programming
- Threads, AsyncTasks, and `Executors`
- The new way to do it – coroutines and flows

By the end of this chapter, you will have a basic understanding of asynchronous programming, and know how to do it in Android using threads, AsyncTasks, and `Executors`. Finally, you will discover Kotlin coroutines and flows as these are the recommended ways of doing asynchronous programming in Android.

Technical requirements

You will need to download and install the latest version of Android Studio. You can find the latest version at `https://developer.android.com/studio`. For an optimal learning experience, a computer with the following specifications is recommended: Intel Core i5 or equivalent or higher, 4 GB RAM minimum, and 4 GB available space.

The code examples for this book can be found on GitHub at https: `github.com/PacktPublishing/` `Simplifying-Android-Development-with-Coroutines-and-Flows`.

Understanding asynchronous programming

In this section, we will start by looking at asynchronous programming. Asynchronous programming is a programming method that allows work to be done independently of the main application thread.

A normal program will run sequentially. It will perform one task and move to the next task after the previous one has finished. For simple operations, this is fine. However, there are some tasks that might take a long time to finish, such as the following:

- Fetching data from or saving data to a database
- Getting, adding, or updating data to a network
- Processing text, images, videos, or other files
- Complicated computations

The app will look frozen and unresponsive to the users while it is performing these tasks. They won't be able to do anything else in the app until the tasks are finished.

Asynchronous programming solves this problem. You can run a task that may be processed indefinitely on a background thread (in parallel to the main thread) without freezing the app. This will allow the users to still interact with the app or the UI while the original task is running. When the task has finished or if an error was encountered, you can then inform the user using the main thread.

A visual representation of asynchronous programming is shown in the following figure:

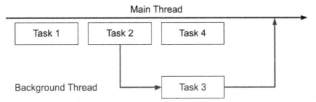

Figure 1.1 – Asynchronous programming

Task 1 and **Task 2** are running on the main thread. **Task 2** starts **Task 3** on the background thread. While **Task 3** is running, the main thread can continue to perform other tasks, such as **Task 4**. After **Task 3** is done, it will return to the main thread.

Asynchronous programming is an important skill for developers to have, especially for mobile app development. Mobile devices have limited capabilities and not all locations have a stable network connection.

In Android, if you run a task on the main thread and it takes too long, the app can become unresponsive or look frozen. The app can also crash unexpectedly. You will likely get an **Application Not Responding** (**ANR**) error, as shown in the following screenshot:

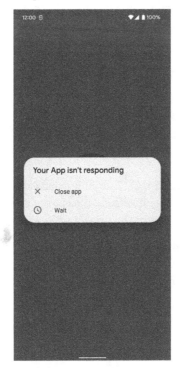

Figure 1.2 – An ANR dialog

Starting with Android 3.0 (Honeycomb), running a network operation on the main thread will cause `android.os.NetworkOnMainThreadException`, which will crash your app.

ANR dialogs and crashes can annoy your users. If they happen all the time, they might stop using your app altogether and choose another app. To prevent them in your app, you must run tasks that can take a long period of time on the background thread.

In this section, you revisited the concept of asynchronous programming and how you can use it to run long-running tasks without freezing the app. You will explore various approaches for using asynchronous programming in Android in the next section.

Exploring threads, AsyncTasks, and Executors

There are many ways you can run tasks on the background thread in Android. In this section, you are going to explore various ways of doing asynchronous programming in Android, including using threads, AsyncTask, and `Executors`. You will learn how to start a task on the background thread and then update the main thread with the result.

Threads

A thread is a unit of execution that runs code concurrently. In Android, the UI thread is the main thread. You can perform a task on another thread by using the `java.lang.Thread` class:

```
private fun fetchTextWithThread() {
  Thread {
        // get text from network
        val text = getTextFromNetwork()
  }.start()
}
```

To run the thread, call `Thread.start()`. Everything that is inside the braces will be performed on another thread. You can do any operation here, except updating the UI, as you will encounter `NetworkOnMainThreadException`.

To update the UI, such as displaying the text fetched in a `TextView` from the network, you would need to use `Activity.runOnUiThread()`. The code inside `runOnUIThread` will be executed in the main thread, as follows:

```
private fun fetchTextWithThread() {
  Thread {
            // get text from network
```

```
            val text = getTextFromNetwork()

    runOnUiThread {
        // Display on UI
        displayText(text)
    }
  }.start()
}
```

runOnUIThread will perform the displayText(text) function on the main UI thread.

If you are not starting the thread from an activity, you can use handlers instead of runOnUiThread to update the UI, as seen in *Figure 1.3*:

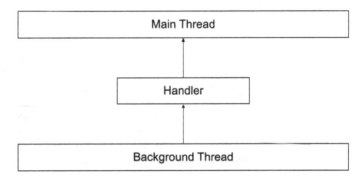

Figure 1.3 – Threads and a handler

A handler (android.os.Handler) allows you to communicate between threads, such as from the background thread to the main thread, as shown in the preceding figure. You can pass a looper into the Handler constructor to specify the thread where the task will be run. A looper is an object that runs the messages in the thread's queue.

To attach the handler to the main thread, you should use Looper.getMainLooper(), like in the following example:

```
private fun fetchTextWithThreadAndHandler() {
  Thread {
    // get text from network
        val text = getTextFromNetwork()

    Handler(Looper.getMainLooper()).post {
      // Display on UI
```

```
        displayText(text)
    }
  }.start()
}
```

`Handler(Looper.getMainLooper())` creates a handler tied to the main thread and posts the `displayText()` runnable function on the main thread.

The `Handler.post (Runnable)` function enqueues the runnable function to be executed on the specified thread. Other variants of the post function include `postAtTime (Runnable)` and `postDelayed (Runnable, uptimeMillis)`.

Alternatively, you can also send an `android.os.Message` object with your handler, as shown in *Figure 1.4*:

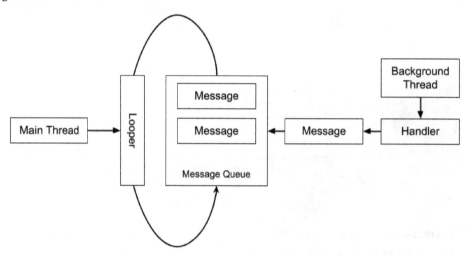

Figure 1.4 – Threads, handlers, and messages

A thread's handler allows you to send a message to the thread's message queue. The handler's looper will execute the messages in the queue.

To include the actual messages you want to send in your Message object, you can use `setData (Bundle)` to pass a single bundle of data. You can also use the public fields of the message `class (arg1, arg2, and what` for integer values, and `obj` for an object value).

You must then create a subclass of Handler and override the `handleMessage (Message)` function. There, you can then get the data from the message and process it in the handler's thread.

You can use the following functions to send a message: sendMessage(Message), sendMessageAtTime(Message, uptimeMillis), and sendMessageDelayed(Message, delayMillis). The following code shows the use of the sendMessage function to send a message with a data bundle:

```
private val key = "key"

private val messageHandler = object :
    Handler(Looper.getMainLooper()) {
      override fun handleMessage(message: Message) {
      val bundle = message.data
      val text = bundle.getString(key, "")
      //Display text
      displayText(text)
    }
}

private fun fetchTextWithHandlerMessage() {
    Thread {
        // get text from network
        val text = getTextFromNetwork()
        val message = handler.obtainMessage()

        val bundle = Bundle()
        bundle.putString(key, text)

        message.data = bundle
        messageHandler.sendMessage(message)
    }.start()
}
```

Here, fetchTextWithHandlerMessage() gets the text from the network in a background thread. It then creates a message with a bundle object containing a string with a key of key to send that text. The handler can then, through the handleMessage() function, get the message's bundle and get the string from the bundle using the same key.

You can also send empty messages with an integer value (the what) that you can use in your handleMessage function to identify what message was received. These send empty functions are sendEmptyMessage(int), sendEmptyMessageAtTime(int, long), and sendEmptyMessageDelayed(int, long).

This example uses 0 and 1 as values for what ("what" is a field of the Message class that is a user-defined message code so that the recipient can identify what this message is about): 1 for the case when the background task succeeded and 0 for the failure case:

```kotlin
private val emptymesageHandler = object :
  Handler(Looper.getMainLooper()) {
  override fun handleMessage(message: Message) {
    if (message.what == 1) {
      //Update UI
    } else {
      //Show Error
    }
  }
}


private fun fetchTextWithEmptyMessage() {
  Thread {
    // get text from network
...
    if (failed) {
      emptyMessageHandler.sendEmptyMessage(0)
    } else {
      emptyMessageHandler.sendEmptyMessage(1)
    }
  }.start()
}
```

In the preceding code snippet, the background thread fetches the text from the network. It then sends an empty message of 1 if the operation succeeded and 0 if not. The handler, through the handleMessage() function, gets the what integer value of the message, which corresponds to the 0 or 1 empty message. Depending on this value, it can either update the UI or show an error to the main thread.

Using threads and handlers works for background processing, but they have the following disadvantages:

- Every time you need to run a task in the background, you should create a new thread and use `runOnUiThread` or a new handler to post back to the main thread.

- Creating threads can consume a lot of memory and resources.

- It can also slow down your app.

- Multiple threads make your code harder to debug and test.

- Code can become complicated to read and maintain.

Using threads makes it difficult to handle exceptions, which can lead to crashes.

As a thread is a low-level API for asynchronous programming, it is better to use the ones that are built on top of threads, such as executors and, until it was deprecated, `AsyncTask`. You can avoid it altogether by using Kotlin coroutines, which you will learn more about later in this chapter.

In the next section, you will explore callbacks, another approach to asynchronous Android programming.

Callbacks

Another common approach to asynchronous programming in Android is using callbacks. A callback is a function that will be run when the asynchronous code has finished executing. Some libraries offer callback functions that developers can use in their projects.

The following is a simple example of a callback:

```
private fun fetchTextWithCallback() {
  fetchTextWithCallback { text ->
    //display text
    displayText(text)
    }
}


fun fetchTextWithCallback(onSuccess: (String) -> Unit) {
    Thread {
        val text = getTextFromNetwork()
        onSuccess(text)
    }.start()
}
```

In the preceding example, after fetching the text in the background, the `onSuccess` callback will be called and will display the text on the UI thread.

Callbacks work fine for simple asynchronous tasks. They can, however, become complicated easily, especially when nesting callback functions and handling errors. This makes it hard to read and test. You can avoid this by avoiding nesting callbacks and splitting functions into subfunctions. It is better to use coroutines, which you will learn more about shortly in this chapter.

AsyncTask

AsyncTask has been the go-to class for running background tasks in Android. It makes it easier to do background processing and post data to the main thread. With `AsyncTask`, you don't have to manually handle threads.

To use `AsyncTask`, you have to create a subclass of it with three generic types:

```
AsyncTask<Params?, Progress?, Result?>()
```

These types are as follows:

- `Params`: This is the type of input for `AsyncTask` or is void if there's no input needed.
- `Progress`: This argument is used to specify the progress of the background operation or Void if there's no need to track the progress.
- `Result`: This is the type of output of `AsyncTask` or is void if there's no output to be displayed.

For example, if you are going to create `AsyncTask` to download text from a specific endpoint, your `Params` will be the URL (`String`) and `Result` will be the text output (`String`). If you want to track the percentage of time remaining to download the text, you can use `Integer` for `Progress`. Your class declaration would look like this:

```
class DownloadTextAsyncTask : AsyncTask<String, Integer,
  String>()
```

You can then start `AsyncTask` with the following code:

```
DownloadTextAsyncTask().execute("https://example.com")
```

`AsyncTask` has four events that you can override for your background processing:

- `doInBackground`: This event specifies the actual task that will be run in the background, such as fetching/saving data to a remote server. This is the only event that you are required to override.

- onPostExecute: This event specifies the tasks that will be run in the UI thread after the background operation finishes, such as displaying the result.

- onPreExecute: This event runs on the UI thread before doing the actual task, usually displaying a progress loading indicator.

- onProgressUpdate: This event runs in the UI thread to denote progress on the background process, such as displaying the amount of time remaining to finish the task.

The diagram in *Figure 1.5* visualizes these AsyncTask events and in what threads they are run:

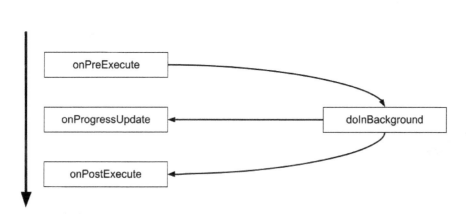

Figure 1.5 – AsyncTask events in main and background threads

The onPreExecute, onProgressUpdate, and onPostExecute functions will run on the main thread, while doInBackground executes on the background thread.

Coming back to our example, your DownloadTextAsync class could look like the following:

```
class DownloadTextAsyncTask : AsyncTask<String, Void,
  String>() {
        override fun doInBackground(vararg params:
          String?): String? {
            valtext = getTextFromNetwork(params[0] ?: "")
            //get text from network
            return text
        }
```

```
        override fun onPostExecute(result: String?) {
            //Display on UI
        }
    }
```

In `DownloadTextAsync`, `doInBackground` fetches the text from the network and returns it as a string. `onPostExecute` will then be called with that string that can be displayed in the UI thread.

`AsyncTask` can cause context leaks, missed callbacks, or crashes on configuration changes. For example, if you rotate the screen, the activity will be recreated and another `AsyncTask` instance can be created. The original instance won't be automatically canceled and when it finishes and returns to `onPostExecute()`, the original activity is already gone.

Using `AsyncTask` also makes your code more complicated and less readable. As of Android 11, `AsyncTask` has been deprecated. It is recommended to use `java.util.concurrent` or Kotlin coroutines instead.

In the next section, you will explore one of the `java.util.concurrent` classes for asynchronous programming, `Executors`.

Executors

One of the classes in the `java.util.concurrent` package that you can use for asynchronous programming is `java.util.concurrent.Executor`. An executor is a high-level Java API for managing threads. It is an interface that has a single function, `execute(Runnable)`, for performing tasks.

To create an executor, you can use the utility methods from the `java.util.concurrent.Executors` class. `Executors.newSingleThreadExecutor()` creates an executor with a single thread.

Your asynchronous code with `Executor` will look like the following:

```
val handler = Handler(Looper.getMainLooper())

private fun fetchTextWithExecutor() {
  val executor = Executors.newSingleThreadExecutor()
  executor.execute {
    // get text from network
            val text = getTextFromNetwork()
```

```
    handler.post {
      // Display on UI
    }
  }
}
```

The handler with `Looper.getMainLooper()` allows you to communicate back to the main thread so you can update the UI after your background task has been done.

`ExecutorService` is an executor that can do more than just `execute(Runnable)`. One of its subclasses is `ThreadPoolExecutor`, an `ExecutorService` class that implements a thread pool that you can customize.

`ExecutorService` has `submit(Runnable)` and `submit(Callable)` functions, which can execute a background task. They both return a `Future` object that represents the result.

The `Future` object has two functions you can use, `Future.isDone()` to check whether the executor has finished the task and `Future.get()` to get the results of the task, as follows:

```
val handler = Handler(Looper.getMainLooper()

private fun fetchTextWithExecutorService() {
  val executor = Executors.newSingleThreadExecutor()
  val future = executor.submit {
    displayText(getTextFromNetwork())
  }
  ...

  val result = future.get()
}
```

In the preceding code, the executor created with a new single thread executor was used to submit the runnable function to get and display text from the network. The `submit` function returns a `Future` object, which you can later use to fetch the result with `Future.get()`.

In this section, you learned some of the methods that you can use for asynchronous programming in Android. While they do work and you can still use them (except for the now-deprecated `AsyncTask`), nowadays, they are not the best method to use moving forward.

In the next section, you will learn the new, recommended way of asynchronous programming in Android: using Kotlin coroutines and flows.

The new way to do it – coroutines and flows

In this section, you will learn about the recommended approach for Android asynchronous programming: using coroutines and flows. Coroutines is a Kotlin library you can use in Android to perform asynchronous tasks. Coroutines is a library for managing background tasks that return a single value. Flows are built on top of coroutines that can return multiple values.

Kotlin coroutines

Coroutines is a Kotlin library for managing background tasks, such as making network calls and accessing files or databases, or performing long-running background tasks. Using Kotlin coroutines is Google's official recommendation for asynchronous programming on Android. Their Android Jetpack libraries, such as Lifecycle, WorkManager, and Room-KTX, now include support for coroutines. Other Android libraries, such as Retrofit, Ktor, and Coil, provide first-class support for Kotlin coroutines.

With Kotlin coroutines, you can write your code in a sequential way. A long-running task can be made into a suspend function. A suspend function is a function that can perform its task by suspending the thread without blocking it, so the thread can still run other tasks. When the suspending function is done, the current thread will resume execution. This makes the code easier to read, debug, and test. Coroutines follow a principle of structured concurrency.

You can add coroutines to your Android project by adding the following lines to your app/build. gradle file dependencies:

```
implementation "org.jetbrains.kotlinx:kotlinx-coroutines-
    core:1.6.0"
implementation "org.jetbrains.kotlinx:kotlinx-coroutines-
    android:1.6.0"
```

kotlinx-coroutines-core is the main library for Kotlin coroutines, while kotlinx-coroutines-android adds support for the main Android thread (Dispatchers.Main).

To mark a function as a suspending function, you can add the suspend keyword to it; for example, here we have a function that calls the fetchText() function, which retrieves text from an endpoint and then displays it in the UI thread:

```
fun fetchText(): String {
    ...
}
```

You can make the fetchText() function a suspending function by prefixing the suspend keyword, as follows:

```
suspend fun fetchText(): String { ... }
```

Then, you can create a coroutine that will call the `fetchText()` suspending function and display the list, as follows:

```
lifecycleScope.launch(Dispatchers.IO) {
    val fetchedText = fetchText()
    withContext(Dispatchers.Main) {
      displayText(fetchedText)
    }
}
```

`lifecycleScope` is the scope with which the coroutine will run. `launch` creates a coroutine to run in `Dispatchers.IO`, which is a thread for I/O or network operations.

The `fetchText()` function will suspend the coroutine before it starts the network request. While the coroutine is suspended, the main thread can do other work.

After getting the text, it will resume the coroutine. `withContext(Dispatchers.Main)` will switch the coroutine context to the main thread, where the `displayText(text)` function will be executed (`Dispatchers.Main`).

In Android Studio, the **Editor** window identifies the `suspend` function calls in your code with a gutter icon next to the line number. As shown in lines **13** and **15** in *Figure 1.6*, the `fetchText()` and `withContext()` lines have the `suspend` function call gutter icon:

```
MainActivity.kt
1     package com.example.yourapp
2
3     import ...
6
7     class MainActivity : AppCompatActivity() {
8         override fun onCreate(savedInstanceState: Bundle?) {
9             super.onCreate(savedInstanceState)
10            setContentView(R.layout.activity_main)
11
12            CoroutineScope(Dispatchers.IO).launch {   this: CoroutineScope
13                val text = fetchText()
14
15                withContext(Dispatchers.Main) {   this: CoroutineScope
16                    displayText(text)
17                }
18            }
19        }
20
21        private suspend fun fetchText(): String {...}
25
26        private fun displayText(text: String) {...}
29    }
30
```

Figure 1.6 – Android Studio suspend function call gutter icon

You can learn more about Kotlin coroutines in *Chapter 2, Understanding Kotlin Coroutines*.

In the next section, you will learn about Kotlin Flows, built on top of coroutines, which can return multiple sequences of values.

Kotlin Flows

Flow is a new Kotlin asynchronous stream library that is built on top of Kotlin coroutines. A flow can emit multiple values instead of a single value and over a period of time. Kotlin Flow is ideal to use when you need to return multiple values asynchronously, such as automatic updates from your data source.

Flow is now used in Jetpack libraries such as Room-KTX and Android developers are already using Flow in their applications.

To use Kotlin Flows in your Android project, you have to add coroutines. An easy way to create a flow of objects is to use the `flow{ }` builder. With the `flow{ }` builder function, you can add values to the stream by calling emit.

Let's say in your Android app you have a `getTextFromNetwork` function that fetches text from a network endpoint and returns it as a `String` object:

```
fun getTextFromNetwork(): String { ... }
```

If we want to create a flow of each word of the text, we can do it with the following code:

```
private fun getWords(): Flow<String> = flow {
  getTextFromNetwork().split(" ").forEach {
    delay(1_000)
    emit(it)
  }
}
```

Flow does not run or emit values until the flow is collected with any terminal operators, such as `collect`, `launchIn`, or `single`. You can use the `collect()` function to start the flow and process each value, as follows:

```
private suspend fun displayWords() {
        getWords().collect {
        Log.d("flow", it)
        }
}
```

A visual representation of this flow is shown in the following figure:

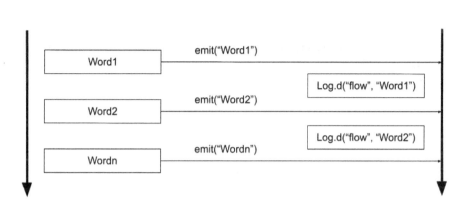

Figure 1.7 – Kotlin Flow visual representation

As you can see in *Figure 1.7*, as soon as the `getWords()` flow emits a string, the `displayWords` function collects the string and displays it immediately on the logs.

You will learn more about Kotlin Flows in *Chapter 5, Using Kotlin Flows*.

In this section, you learned about Kotlin coroutines and flows, the recommended way to carry out asynchronous programming in Android. Coroutines is a Kotlin library for managing long-running tasks in the background. Flow is a new Kotlin asynchronous stream library, built on top of coroutines, that can emit multiple values over a period of time.

Summary

In this chapter, you revisited the concept of asynchronous programming. We learned that asynchronous programming helps you execute long-running tasks in the background without freezing the app and annoying your users.

You then learned about various ways you can do asynchronous programming in Android, including with threads, AsyncTask, and `Executors`. We also learned that they allow you to perform tasks in the background and update the main thread. AsyncTask is already deprecated, and threads and `Executors` are not the best ways to carry out asynchronous programming in Android.

Finally, you were introduced to the new, recommended way to carry out asynchronous programming in Android: with Kotlin's Coroutines and Flow. We learned that Coroutines is a Kotlin library that you can use to easily perform asynchronous, non-blocking, and long-running tasks in the background. Flow, built on top of Coroutines, allows you to handle functions that return multiple values over time.

In the next chapter, you will dive deeper into Kotlin coroutines and learn how to use them in your Android project.

Further reading

This book assumes that you have experience and skills in Android development with Kotlin. If you would like to learn more about this, you can read the book *How to Build Android Apps with Kotlin* (*Packt Publishing, 2021, ISBN 9781838984113*).

2
Understanding Kotlin Coroutines

In the previous chapter, you revisited the concept of asynchronous programming and how it helps you execute long-running tasks in the background without freezing the app and annoying your app's users. You learned how to carry out asynchronous programming with Threads, AsyncTasks, and Executors. Finally, you were introduced to the new way of doing it on Android: Kotlin Coroutines and Flows.

Coroutines is a Kotlin library for multithreading and asynchronous programming, such as making network calls and accessing files or databases. Kotlin Coroutines is Google's official recommendation for asynchronous programming on Android. Android Jetpack libraries, such as ViewModel, Lifecycle, WorkManager, and Room, include support for Kotlin Coroutines. Third-party Android libraries, such as Retrofit, now provide support for Kotlin Coroutines.

In this chapter, we will dive deep into Kotlin Coroutines. You will learn how to use coroutines to carry out asynchronous programming in Android with simple code. You will also learn how to create coroutines in your Android app. Then, we will discuss other building blocks of coroutines, such as builders, scopes, dispatchers, contexts, and jobs.

In this chapter, we're going to cover the following topics:

- Creating coroutines in Android
- Exploring coroutine builders, scopes, and dispatchers
- Understanding coroutine contexts and jobs
- Exercise – using coroutines in an Android app

By the end of this chapter, you will have an understanding of using Kotlin coroutines. You will be able to add coroutines for various cases in your Android apps. You will also understand the basic building blocks of Coroutines: builders, scopes, dispatchers, contexts, and jobs.

Technical requirements

For this chapter, you will need to download and install the latest version of Android Studio. You can find the latest version at `https://developer.android.com/studio`. For an optimal learning experience, a computer with the following specifications is recommended: Intel Core i5 or equivalent or higher, 4 GB of RAM minimum, and 4 GB available space.

The code examples for this chapter can be found on GitHub at `https://github.com/PacktPublishing/Simplifying-Android-Development-with-Coroutines-and-Flows/tree/main/Chapter02`.

Creating coroutines in Android

In this section, we will start by looking at how to create coroutines in Android. Coroutines provide an easy way to write asynchronous code with Kotlin's standard functions. You can use coroutines when making a network call or when fetching data from or saving data to a local database.

A simple coroutine looks as follows:

```
CoroutineScope(Dispatchers.IO).launch {
    performTask()
    ...
}
```

It has four parts: `CoroutineScope`, `Dispatchers`, `launch`, and the lambda function that will be executed by the coroutine. An instance of `CoroutineScope` was created for the coroutine's scope. `Dispatchers.IO` is the dispatcher that will specify that this coroutine will run on the I/O dispatcher, the one usually used for **input/output (I/O)** operations such as networking, database, and file processing. `launch` is the coroutine builder that creates the coroutine. We will explore these components in detail later in this chapter.

The following diagram summarizes these parts of a coroutine:

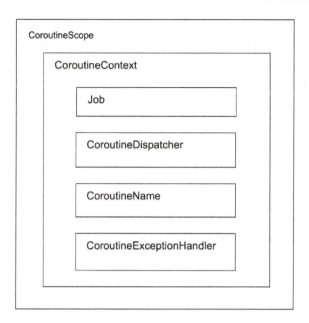

Figure 2.1 – Parts of a coroutine

In Android Studio, the **Editor** window identifies the suspending function calls in your code with a gutter icon next to the line number. As shown on line **16** in the following screenshot, the `performTask()` call has the suspend function call gutter icon next to it:

```
MainActivity.kt ×
1      package com.example.yourapp
2
3      import ...
9
10     class MainActivity : AppCompatActivity() {
11         override fun onCreate(savedInstanceState: Bundle?) {
12             super.onCreate(savedInstanceState)
13             setContentView(R.layout.activity_main)
14
15             CoroutineScope(Dispatchers.IO).launch {   this: CoroutineScope
16                 performTask()
17             }
18         }
19
20         private suspend fun performTask() {...}
25     }
26
```

Figure 2.2 – Android Studio suspend function call gutter icon

Let's say you have an Android application that displays the list of movies that are currently playing in cinemas. So, let's look at the ways you can use the suspend function and add coroutines to the project.

If you're using Retrofit 2.6.0 or above, you can mark the endpoint function as a suspending function with suspend, as follows:

```
@GET("movie/now_playing")
suspend fun getMovies() : List<Movies>
```

Then, you can create a coroutine that will call the getMovies suspending function and display the list:

```
CoroutineScope(Dispatchers.IO).launch {
    val movies = movieService.getMovies()
    withContext(Dispatchers.Main) {
        displayMovies(movies)
    }
}
```

This will create a coroutine that fetches the movies in the background. The withContext call will change the context of the coroutine to use Dispatchers.Main to display the fetched movies in the main thread.

If you are using Room-KTX 2.1 or above, you can add the suspend keyword to your **Data Access Object (DAO)** functions so that the query or operation can be executed on the background thread and the result will be posted on the main thread. The following is an example:

```
@Dao
interface MovieDao {
    @Query("SELECT * from movies")
    suspend fun getMovies(): List<Movies>


    . . .

}
```

This will make the getMovies query a suspending function. When you call this function, Room-KTX internally executes the query on a background thread. The results can be displayed on the main thread without it freezing your app.

When you create a coroutine inside another coroutine, the new coroutine becomes the child of the original coroutine. The original coroutine becomes the parent of the new coroutine. This can be seen in the following code:

```
CoroutineScope(Dispatchers.IO).launch {
    performTask1()
    launch {
        performTask2()
    }
    ...
}
```

The second coroutine that was launched with `performTask2` was created using the `Coroutine Scope` of the parent coroutine.

In this section, you explored how to add coroutines to your Android project and learned how to create coroutines for your app. In the next section, you will explore some of the building blocks of coroutines: builders, scopes, and dispatchers.

Exploring coroutine builders, scopes, and dispatchers

In this section, you will learn how to use coroutine builders and explore coroutine scopes and dispatchers. Coroutine builders are functions that are used to create coroutines. Coroutine scope is the scope with which the coroutines run. Dispatchers specify in what thread the coroutine will run.

Coroutine builders

In the previous section, you created a coroutine with `launch`. However, there are other ways to create coroutines. Coroutine builders are the functions that you can use to create coroutines. To create a coroutine, you can use the following Kotlin coroutine builders:

- `launch`
- `async`
- `runBlocking`

`async` and `launch` need to be started on a coroutine scope. Meanwhile, `runBlocking` doesn't need to be started from a coroutine scope.

The `launch` keyword creates a coroutine and doesn't return a value. Instead, it returns a `Job` object that represents the coroutine.

The `launch` coroutine builder is ideal to use when you want to run a task and then forget about it (this means you are not waiting for the result of the operation). Here's an example of using the `launch` coroutine builder:

```
class MainActivity : AppCompatActivity() {

    val scope = MainScope()

    override fun onCreate(savedInstanceState: Bundle?) {
        super.onCreate(savedInstanceState)
        setContentView(R.layout.activity_main)

        val progressBar =
          findViewById<ProgressBar>(R.id.progressBar)

        scope.launch {
            delay(1_000)
            progressBar.isVisible = true
        }
    }
}
```

Once the activity has been created, a coroutine will be launched. This coroutine will call the `delay` suspending function to delay the coroutine for a second, resume, and display the progress bar; then, it's done.

On the other hand, the `async` builder is similar to `launch` but it returns a value: a `Deferred` object. Later, you can get this value with the `await` function. The `async` builder should be used when you want to execute a task and want to get the output of said task. An example of using the `async` coroutine builder is as follows:

```
class MainActivity : AppCompatActivity() {

    val scope = MainScope()

    override fun onCreate(savedInstanceState: Bundle?) {
        super.onCreate(savedInstanceState)
        setContentView(R.layout.activity_main)
```

```
        val textView =
          findViewById<TextView>(R.id.textView)

        scope.launch {
            val text = async {
                getText()
            }

            delay(1_000)
            textView.text = text.await()

        }
    }
}
```

Here, a coroutine was started with `async` to call the `getText` function. This will return a deferred object called `text`. There will be a delay of 1 second and then the actual value from `text` will be called with `text.await()`, which will be set as the text for `textView`. With `async`, two tasks can be computed in parallel.

`runBlocking` starts a new coroutine and blocks the current thread until the task has been executed. This is useful for cases when you need to block the thread. Creating unit tests is one of these cases:

```
class MainActivity : AppCompatActivity() {

    override fun onCreate(savedInstanceState: Bundle?) {
        super.onCreate(savedInstanceState)
        setContentView(R.layout.activity_main)

        val progressBar =
          findViewById<ProgressBar>(R.id.progressBar)

        runBlocking {
            delay(2_000)
            progressBar.isVisible = true
        }
    }
}
```

In the preceding code, the `runBlocking` code will create a coroutine and block the thread. After a delay of 2,000 milliseconds (2 seconds), it will display the progress bar.

In this section, you explored how to use coroutine builders to create coroutines. You also learned about the `async`, `launch`, and `runBlocking` coroutine builders.

In the next section, you will explore coroutine scopes.

Coroutine scopes

`CoroutineScope` is the scope with which the coroutine will run. It defines the life cycle of the coroutines created from it, from its start to its end. If you cancel a scope, it will cancel all the coroutines it created. Coroutines follow the principle of structured concurrency – that is, a mechanism that provides the structure of a coroutine.

The `launch` and `async` coroutine builders are extension functions from `CoroutineScope` for creating coroutines.

For example, let's say we have created a coroutine using `MainScope`:

```
class MainActivity : AppCompatActivity() {

    val scope = MainScope()

    override fun onCreate(savedInstanceState: Bundle?) {
        super.onCreate(savedInstanceState)
        setContentView(R.layout.activity_main)

        val progressBar =
          findViewById<ProgressBar>(R.id.progressBar)

        scope.launch {
            progressBar.isVisible = true
        }
    }
}
```

This launches a coroutine using `MainScope` to display the progress bar.

`MainScope` is the main `CoroutineScope` for the main thread, which uses `Dispatchers.Main` for its coroutine. It is normally used for creating coroutines that will update the user interface.

You can also create a CoroutineScope instead of using MainScope by creating one with the CoroutineScope factory function. The CoroutineScope function requires you to pass in a coroutine context. CoroutineContext is a collection of elements for the coroutines that specify how the coroutine should run.

You passed a dispatcher and both a dispatcher and a job in the previous examples for the coroutine context. Dispatchers and jobs are coroutine context elements. You will learn more about coroutine contexts later in this chapter.

Your CoroutineScope must have a job and a way for the coroutine to be canceled, such as when Activity, Fragment, or ViewModel has been closed.

In the next section, we will look at a built-in coroutine scope called lifecycleScope, which is part of Jetpack's Lifecycle library.

lifecycleScope

lifecycleScope is a CoroutineScope from Jetpack's Lifecycle library that you can use to create coroutines. It is tied to the Lifecycle object (similar to your activity or fragment) and is automatically canceled when the life cycle is destroyed. Thus, you no longer need to manually cancel them.

lifecycleScope simplifies how scopes are created, how jobs are handled, and how they can be canceled within your activity or fragment. A lifecycleScope uses Dispatchers.Main. immediate for its dispatcher and a SupervisorJob for its job, such as viewModelScope.

To use lifecycleScope, you must add the following line to your app/build.gradle file dependencies:

```
implementation "androidx.lifecycle:lifecycle-runtime-ktx:2.4.1"
```

An example of lifeCycleScope is as follows:

```
class MainActivity : AppCompatActivity() {

    override fun onCreate(savedInstanceState: Bundle?) {
        super.onCreate(savedInstanceState)
        setContentView(R.layout.activity_main)

        val progressBar =
            findViewById<ProgressBar>(R.id.progressBar)

        lifecycleScope.launch {
```

```
                progressBar.isVisible = true
        }
    }
}
```

When the activity is created, it launches a coroutine from `lifecycleScope` to display the progress bar.

To change the dispatcher that the coroutine will use, you can pass in a dispatcher when using the `launch` and `async` coroutine builders:

```
lifecycleScope.launch(Dispatchers.IO) { ... }
```

This will use the `Dispatchers.IO` dispatcher instead of the `lifecycleScope` object's default `Dispatchers.Main.immediate` for the coroutine that was launched.

Aside from `launch`, `lifecycleScope` has additional coroutine builders, depending on the life cycle's state:

- `launchWhenCreated`
- `launchWhenStarted`
- `launchWhenResumed`

As the name suggests, `launchWhenCreated` launches the coroutine when the life cycle is created, `launchWhenStarted` launches the coroutine when the life cycle is started, and `launchWhenResumed` launches the coroutine when the life cycle goes back to the **Resumed** state.

In the next section, we will look at a built-in `CoroutineScope` from `ViewModel` called `viewModelScope`.

viewModelScope

`viewModelScope` is the ViewModel's default `CoroutineScope` for creating coroutines. It is ideal to use if you need to do a long-running task from `ViewModel`. This scope and all running jobs are automatically canceled when `ViewModel` is cleared (that is, when `onCleared` is invoked).

`viewModelScope` simplifies the creation of `Scope`, handling the job, and canceling within `ViewModel`. A `viewModelScope` uses `Dispatchers.Main.immediate` for its dispatcher and uses a `SupervisorJob` for the job. A `SupervisorJob` is a special version of `Job` that allows its children to fail independently of each other.

To use `viewModelScope`, you must add the following line to your `app/build.gradle` file dependencies:

```
implementation "androidx.lifecycle:lifecycle-viewmodel-
   ktx:2.4.1"
```

You can use `viewModelScope` like so:

```
class MovieViewModel: ViewModel() {
    init {
        viewModelScope.launch {
            fetchMovies()
        }
    }
}
```

This launches a coroutine from `viewModelScope` that will be used to run the `fetchMovies()` function.

To change the dispatcher that the coroutine will use, you can pass in a dispatcher when using the `launch` and `async` coroutine builders:

```
viewModelScope.launch (Dispatchers.IO) { ... }
```

This will use `Dispatchers.IO` for the coroutine, instead of viewModelScope's default of `Dispatchers.Main`.

coroutineScope{} and supervisorScope{}

The `coroutineScope{}` suspending builder allows you to create a `CoroutineScope` with the coroutine context from its outer scope. This calls the code block inside and does not complete until everything is done.

You can use a `coroutineScope{}` builder like so:

```
private suspend fun fetchAndDisplay() = coroutineScope {
        launch {
            val movies = fetchMovies()
            displayMovies(movies)
        }
```

```
        launch {
            val shows = fetchShows()
            DisplayShows(shows)
        }
    }
```

This will create a coroutine scope that will call the `fetchMovies` function, set its return value to the `movies` object, and then call the `displayMovies` function with `movies`. Another child coroutine will call the `fetchShows` function, set its return value to the `shows` object, and then call the `displayShows` function with `shows`.

When a child coroutine fails, it will cancel the parent coroutine and the sibling coroutines. If you do not want this to happen, you can use `supervisorScope{}` instead of `coroutineScope{}`.

The `supervisorScope{}` builder is similar to the `coroutineScope{}` builder but the coroutine's `Scope` has a `SupervisorJob`. This allows the children of `supervisorScope` to fail independently of each other.

An example of `supervisorScope` is as follows:

```
private suspend fun fetchAndDisplayMovies() =
    supervisorScope {
        launch {
            val movies = fetchMovies()
            displayMovies(movies)
        }
        launch {
            val shows = fetchShows()
            displayShows(shows)
        }

    }
```

This will create a supervisor scope (with a `SupervisorJob`) that will call the `fetchMovies` function. When a child coroutine fails, the parent and sibling coroutines will continue to work and will not be affected.

GlobalScope

GlobalScope is a special CoroutineScope that is not tied to an object or a job. It should only be used in cases when you must run a task or tasks that will always be active while the application is alive. As such, if you want to use GlobalScope, you must annotate the call with @ OptIn(DelicateCoroutinesApi::class).

For all other cases in Android, it is recommended to use viewModelScope, lifecycleScope, or a custom coroutine scope.

Test scopes

Kotlin has a kotlinx-coroutines-test library for testing coroutines. This testing library includes a special coroutine scope that you can use to create tests for your coroutines. You will learn more about testing coroutines in *Chapter 4, Testing Kotlin Coroutines*.

In this section, you learned about CoroutineScope, as well as about MainScope and creating coroutine scopes with the CoroutineScope function. You also learned about built-in scopes such as viewModelScope and lifecycleScope.

In the next section, you will learn about coroutine dispatchers.

Coroutine dispatchers

Coroutines have a context, which includes the coroutine dispatcher. Dispatchers specify what thread the coroutine will use to perform the task. The following dispatchers can be used:

- Dispatchers.Main: This is used to run on Android's main thread, usually for updates to the user interface. A special version of Dispatchers.Main, called Dispatchers.Main.immediate, is used to immediately execute the coroutine in the main thread. The viewModelScope and lifecycleScope coroutine scopes use Dispatchers.Main.immediate by default.

- Dispatchers.IO: This is designed for networking operations, and for reading from or writing to files or databases.

- Dispatchers.Default: This is used for CPU-intensive work, such as complicated computations or processing text, images, or videos. If you don't set a dispatcher, Dispatchers.Default will be chosen by default.

- Dispatchers.Unconfined: This is a special dispatcher that is not confined to any specific threads. It executes the coroutine in the current thread and resumes it in whatever thread that is used by the suspending function.

You can set the dispatchers when setting the context in `CoroutineScope` or when using coroutine builders.

When using `MainScope` as the coroutine scope for your coroutine, `Dispatchers.Main` is used by default:

```
MainScope().launch { ... }
```

This coroutine will automatically use `Dispatchers.Main` so that you no longer need to specify it.

If you used a different coroutine scope, you can pass in the dispatcher that will be used by the coroutine:

```
CoroutineScope(Dispatchers.IO).launch {
    fetchMovies()
}
```

The preceding code creates a `CoroutineScope` that will be run using `Dispatchers.IO` for the dispatcher.

You can also pass in a dispatcher when using the `launch` and `async` coroutine builders:

```
viewModelScope.launch(Dispatchers.Default) { ... }
```

This will launch a coroutine using the `Dispatchers.Default` dispatcher.

To change the context of your coroutine, you can use the `withContext` function for the code that you want to use a different thread with. For example, in your suspending function, `getMovies`, which gets movies from your endpoint, you can use `Dispatchers.IO`:

```
suspend fun getMovies(): List<Movies>  {
    withContext(Dispatchers.IO) { ... }
}
```

In the preceding code, the `getMovies` function uses `Dispatchers.IO` to fetch the list of movies from a network endpoint.

In this section, you learned what dispatchers are and what dispatchers you can use, depending on your needs. You also learned how to use `withContext` to change the specific thread the coroutine runs on.

In the next section, you will explore coroutine contexts and jobs.

Understanding coroutine contexts and jobs

In this section, you will learn about coroutine contexts and jobs. Coroutines run in a coroutine context. A job is the context of the coroutine that allows you to manage the coroutine's execution.

Coroutine contexts

Each coroutine runs in a coroutine context. A coroutine context is a collection of elements for the coroutines that specifies how the coroutine should run. A coroutine scope has a default coroutine context; if it's empty, it will have an `EmptyCoroutineContext`.

When you create a `CoroutineScope` or use a coroutine builder, you can pass in a `CoroutineContext`. In the previous examples, we were passing a dispatcher:

```
CoroutineScope(Dispatchers.IO) {
    ...
}


viewModelScope.launch(Dispatchers.Default) { ... }
```

The preceding example shows how to pass a dispatcher in the `CoroutineScope` function or in the coroutine builder.

What you're passing in these functions is a `CoroutineContext`. The following are some of the `CoroutineContext` elements you can use:

- `CoroutineDispatcher`
- `Job`
- `CoroutineName`
- `CoroutineExceptionHandler`

The main `CoroutineContext` elements are dispatchers and jobs. Dispatchers specify the thread where the coroutine runs, while the job of the coroutine allows you to manage the coroutine's task.

Jobs allow you to manage the life cycle of the coroutine, from the creation of the coroutine to the completion of the task. You can use this job to cancel the coroutine itself. You'll learn more about coroutine cancelations in *Chapter 3, Handling Coroutines Cancelations and Exceptions*.

`CoroutineName` is another `CoroutineContext` you can use to set a string to name a coroutine. This name can be useful for debugging purposes. For example, you can add a `CoroutineName` using the following code:

```
val scope = CoroutineScope(Dispatchers.IO)
scope.launch(CoroutineName("IOCoroutine")) {
    performTask()
}
```

This will give the name of `IOCoroutine` to the coroutine that was launched using the `Dispatchers.IO` dispatcher.

As the coroutine context is a collection of elements for the coroutine, you can use operators such as the + symbol to combine context elements to create a new `CoroutineContext`:

```
val context = Dispatchers.Main + Job()
```

`MainScope`, `viewModelScope`, and `lifecycleScope`, for example, use something like the following for the coroutine scope's context:

```
SupervisorJob() + Dispatchers.Main.immediate
```

Another coroutine context element you can use is `CoroutineExceptionHandler`, an element you can use to handle exceptions. You will learn more about `CoroutineExceptionHandler` in *Chapter 3, Handling Coroutines Cancelations and Exceptions*.

In the previous section, you used the `withContext` function to change the dispatcher to specify a different thread to run your coroutine. As the name implies, this changes the coroutine context with the dispatcher, which is a `CoroutineContext` element itself:

```
withContext(Dispatchers.IO) { ... }
```

This changes the coroutine context with a new dispatcher, `Dispatchers.IO`.

In the next section, you will learn about jobs.

Coroutine jobs

A **job** is a `ContextCoroutine` element that you can use for the coroutine context. You can use jobs to manage the coroutine's tasks and its life cycle. Jobs can be canceled or joined together.

The `launch` coroutine builder creates a new job, while the `async` coroutine builder returns a `Deferred<T>` object. `Deferred` is itself a `Job` object – that is, a job that has a result.

To access the job from the coroutine, you can set it to a variable:

```
val job = viewModelScope.launch(Dispatchers.IO) { ... }
```

The `launch` coroutine builder creates a coroutine running in the `Dispatchers.IO` thread and returns a job. A job can have children jobs, making it a parent job. `Job` has a `children` property you can use to get the job's children:

```
val job1 = viewModelScope.launch(Dispatchers.IO) {
    val movies = fetchMovies()
```

```
    val job2 = launch {
        . . .
    }
    . . .
}
```

In this example, job2 becomes a child of job1, which is the parent. This means that job2 will inherit the coroutine context of the parent, though you can also change it.

If a parent job is canceled or failed, its children are also automatically canceled. When a child's job is canceled or failed, its parent will also be canceled.

A SupervisorJob is a special version of a job that allows its children to fail independently of each other.

Using a job also allows you to create a coroutine that you can later start instead of immediately running by default. To do this, you must use CoroutineStart.LAZY as the value of the start parameter in your coroutine builder and assign the result to a Job variable. Later, you can then use the start() function to run the coroutine, as shown here:

```
val lazyJob = viewModelScope.launch (start=CoroutineStart.LAZY)
{
    delay(1_000)
    . . .
}
. . .
lazyJob.start()
```

This will create a lazy coroutine. When you are ready to start it, you can simply call lazyJob.start().

With the Job object, you can also use the join() suspending function to wait for the job to be completed before continuing with another job or task:

```
viewModelScope.launch {
    val job1 = launch {
        showProgressBar()
    }
    . . .
    job1.join()
    . . .
```

```
    val job2 = launch {
        fetchMovies()
    }
}
```

In this example, job1 will be run first and job2 won't be executed until the former job (job1) has finished.

In the next section, you will learn more about the states of coroutine jobs.

Coroutine job states

A job has the following states:

- New
- Active
- Completing
- Completed
- Canceling
- Canceled

These states of a job and its life cycle are summarized in the following diagram:

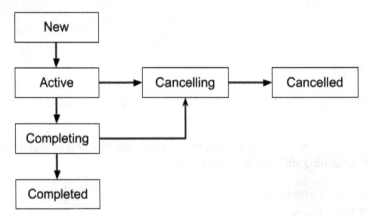

Figure 2.3 – Coroutine job life cycle

When you start a coroutine, a job is created in the **New** state. It then becomes **Active** when the job automatically starts or when you call the start() or join() function. The job is in the **Active** state while the job is running.

Completing a job moves it into the **Completing** state and then into the **Completed** state once its children complete their tasks.

If a job is canceled manually or failed because of an exception, it will go into the **Canceling** state and then into the **Canceled** state once its children complete.

A `Job` object has three properties you can use to check the state of the job:

- `isActive`: This property is `true` when the job is running or completing, and `false` otherwise.

- `isComplete`: This property is `true` when the job has finished its task (canceled or completed), and `false` otherwise.

- `isCancelled`: This property is `true` if the job has been canceled or is being canceled, manually or because of an exception, and `false` otherwise.

You will learn more about jobs and how they are used to cancel coroutines in *Chapter 3, Handling Coroutines Cancelations and Exceptions*.

In this section, you learned about coroutine contexts and jobs. `CoroutineContext` is a collection of coroutine context elements for the coroutines that specifies how the coroutine should run. Examples of `CoroutineContext` elements are dispatchers and jobs. A job is created from a coroutine. You can use it to manage the coroutine's tasks and life cycle.

Now, you will use what you have learned so far to add coroutines to an Android project.

Exercise – using coroutines in an Android app

In this exercise, you will be working with an application that displays movies that are playing now in cinemas. You will be using The Movie Database API version 3 to get the list of movies. Go to `https://developers.themoviedb.org/3` and register for an API key. Once you've done that, follow these steps:

1. Open the `Movie App` project in the `Chapter02` directory in this book's code repository.

2. Open `MovieRepository` and update `apiKey` with the value from The Movie Database API:

   ```
   private val apiKey = "your_api_key_here"
   ```

3. Open the `app/build.gradle` file and add a dependency for `kotlinx-coroutines-android`:

   ```
   implementation 'org.jetbrains.kotlinx:kotlinx-
      coroutines-android:1.6.0'
   ```

This will add the `kotlinx-coroutines-core` and `kotlinx-coroutines-android` libraries to your project, allowing you to use coroutines in your code.

4. Also, add the dependencies for the `ViewModel` extension library:

```
implementation 'androidx.lifecycle:lifecycle-
    viewmodel-ktx:2.4.1'
```

This will add the `ViewModel` KTX library to your project. It includes a `viewModelScope` for `ViewModel`.

5. Open the `MovieViewModel` class, navigate to the `fetchMovies` function, and add the following code:

```
fun fetchMovies() {
    _loading.value = true
    viewModelScope.launch(Dispatchers.IO) {
    }
}
```

This will create a coroutine that will run in `Dispatchers.IO` (on a background thread for network operations). The coroutine will be launched using `viewModelScope`.

6. In the `fetchMovies` coroutine, call the MovieRepository's `fetchMovies` function to fetch the list of movies from The Movie Database API:

```
fun fetchMovies() {
    _loading.value = true
    viewModelScope.launch(Dispatchers.IO) {
        movieRepository.fetchMovies()
        _loading.postValue(false)
    }
}
```

The coroutine will be launched and will call the `fetchMovies` function from `MovieRepository`.

7. Run the application. You will see that the app displays a list of movies (with a poster and a title), as shown in the following screenshot:

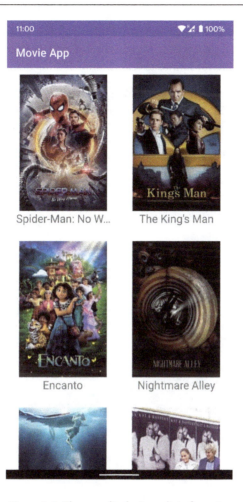

Figure 2.4 –The app displaying a list of movies

In this exercise, you created a coroutine using the ViewModel's `viewModelScope`, used the `launch` coroutine builder, and executed a task to fetch the movies from the repository.

Summary

In this chapter, you learned more about Kotlin coroutines and how you can use them for asynchronous programming in Android.

You learned how to create coroutines with coroutine builders such as `launch`, `async`, and `runBlocking`. Then, you learned about dispatchers and using them to set the thread that the coroutines run on. You also learned about coroutine scopes and built-in scopes such as `viewModelScope` and `lifecycleScope`.

After that, you learned about coroutine contexts and jobs. `CoroutineContext` is the context for the coroutine and includes elements such as dispatchers that the coroutine will run on and a job, which you can use to manage the coroutine's task.

Finally, you completed an exercise where you added a coroutine to an Android project. You used the ViewModel's `viewModelScope` for the coroutine scope, the `launch` coroutine builder, and implemented the coroutine to fetch the list of movies using `Dispatchers.IO`.

In the next chapter, you will learn how to handle coroutine cancelations, timeouts, and exceptions.

3
Handling Coroutine Cancelations and Exceptions

In the previous chapter, you dove deep into Kotlin coroutines and learned how to use them for asynchronous programming in Android with simple code. You learned how to create coroutines with coroutine builders. Finally, you explored coroutine dispatchers, coroutine scopes, coroutine contexts, and jobs.

Coroutines can be canceled when their purpose has been fulfilled or their job has been done. You can also cancel them based on specific instances in your app, such as when you want users to manually stop a task with a tap of a button. Coroutines do not always succeed and can fail; developers must be able to handle these cases so that the app will not crash, and they can inform the users by displaying a toast or snackbar message.

In this chapter, we will start by understanding coroutine cancelation. You will learn how to cancel coroutines and handle cancelations and timeouts for your coroutines. Then, you will learn how to manage failures and exceptions that can happen in your coroutines.

In this chapter, we will cover the following topics:

- Canceling coroutines
- Managing coroutine timeouts
- Catching exceptions in coroutines

By the end of this chapter, you will understand coroutine cancelations and how you can make your coroutines cancelable. You will be able to add and handle timeouts in your coroutines. You will also know how to add code to catch exceptions in your coroutines.

Technical requirements

You will need to download and install the latest version of Android Studio. You can find the latest version at `https://developer.android.com/studio`. For an optimal learning experience, a computer with the following specifications is recommended: Intel Core i5 or equivalent or higher, 4 GB RAM minimum, and 4 GB available space.

The code examples for this chapter can be found on GitHub at `https://github.com/PacktPublishing/Simplifying-Android-Development-with-Coroutines-and-Flows/tree/main/Chapter03`.

Canceling coroutines

In this section, we will start by looking at coroutine cancelations Developers can cancel coroutines in their projects manually or programmatically. You must make sure your application can handle these cancelations.

If your application is doing a long-running operation that is taking longer than expected and you think it could cause a crash, you might want to stop that task. You can also end tasks that are no longer necessary to free up memory and resources, such as when the user moves out of the activity that launched the task or closes the application. Users can also manually discontinue certain operations if you have that feature in your application. Coroutines make it easier for developers to cancel these tasks.

If you are using `viewModelScope` from `ViewModel` or `lifecycleScope` from the Jetpack Lifecycle Kotlin extension libraries, you can easily create coroutines without manually handling the cancelation. When `ViewModel` is cleared, `viewModelScope` is automatically canceled, while `lifecycleScope` is automatically canceled when the life cycle is destroyed. If you created your own coroutine scope, you must add the cancelation yourself.

In the previous chapter, you learned that using coroutine builders such as `launch` returns a **job**. Using this **job** object, you can call the `cancel()` function to cancel the coroutine. Take the following example:

```
class MovieViewModel: ViewModel() {
    init {
        viewModelScope.launch {
            val job = launch {
                fetchMovies()
            }
            ...
            job.cancel()
        }
```

```
        }
    }
```

The `job.cancel()` function will cancel the coroutine launched to call the `fetchMovies()` function.

After canceling the job, you may want to wait for the cancelation to be finished before continuing to the next task to avoid race conditions. You can do that by calling the `join` function after calling the `call` function:

```
class MovieViewModel: ViewModel() {
    init {
        viewModelScope.launch() {
            val job = launch {
                fetchMovies()
            }
            ...
            job.cancel()
            job.join()
            hideProgressBar()
        }
    }
}
```

Adding `job.join()` here would make the code wait for the job to be canceled before doing the next task, which is `hideProgressBar()`.

You can also use the `Job.cancelAndJoin()` extension function, which is the same as calling `cancel` and then the `join` function:

```
class MovieViewModel: ViewModel() {
    init {
        viewModelScope.launch() {
            val job = launch {
                fetchMovies()
            }
            ...
            job.cancelAndJoin()
            hideProgressBar()
        }
```

```
        }
    }
```

The `cancelAndJoin` function simplifies the call to the `cancel` and `join` functions into a single line of code.

Coroutine jobs can have child coroutine jobs. When you cancel a job, its child jobs (if there are any) will also be canceled, recursively.

If your coroutine scope has multiple coroutines and you need to cancel all of them, you can use the `cancel` function from the coroutine scope instead of canceling the jobs one by one. This will cancel all the coroutines in the scope. Here's an example of using the coroutine scope's `cancel` function to cancel coroutines:

```
class MovieViewModel: ViewModel() {
    private val scope = CoroutineScope(Dispatchers.Main +
      Job())
    init {
        scope.launch {
            val job1 = launch {
                fetchMovies()
            }
            val job2 = launch {
                displayLoadingText()
            }
        }
    }


    override fun onCleared() {
        scope.cancel()
    }
}
```

In this example, when `scope.cancel()` is called, it will cancel both the `job1` and `job2` coroutines, which were created in the coroutine `scope` scope.

Using the `cancel` function from the coroutine scope makes it easier to cancel multiple jobs launched with the specified scope. However, the coroutine scope won't be able to launch new coroutines after you called the `cancel` function on it. If you want to cancel the scope's coroutines but still want to create coroutines from the scope later, you can use `scope.coroutineContext.cancelChildren()` instead:

```
class MovieViewModel: ViewModel() {
    private val scope = CoroutineScope(Dispatchers.Main +
      Job())
    init {
        scope.launch() {
            val job1 = launch {
                fetchMovies()
            }

            val job2 = launch {
                displayLoadingText()
            }
        }
    }


    fun cancelAll() {
        scope.coroutineContext.cancelChildren()
    }

    ...


}
```

Calling the `cancelAll` function will cancel all the child jobs in the coroutine context of the scope. You will still be able to use the scope later to create coroutines.

Canceling a coroutine will throw `CancellationException`, a special exception that indicates the coroutine was canceled. This exception will not crash the application. You will learn more about coroutines and exceptions later in this chapter.

You can also pass a subclass of `CancellationException` to the `cancel` function to specify a different cause:

```
class MovieViewModel: ViewModel() {
private lateinit var movieJob: Job

    init {
        movieJob = scope.launch() {
            fetchMovies()
        }
    }


    fun stopFetching() {
        movieJob.cancel(CancellationException("Cancelled by
            user"))
    }

    ...

}
```

This cancels the `movieJob` job with `CancellationException` containing the message `Cancelled by user` as the cause when the user calls the `stopFetching` function.

When you cancel a coroutine, the coroutine's job's state will change to `Cancelling`. It won't automatically go to the `Cancelled` state and cancel the coroutine. The coroutine can continue to run even after the cancelation, unless your coroutine has code that can stop it from running. These states of a job and its life cycle are summarized in the following diagram:

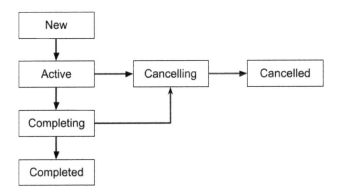

Figure 3.1 – Coroutine job life cycle

Your coroutine code needs to cooperate to be cancelable. The coroutine should handle cancelations as quickly as possible. It must check for cancelations of the coroutine and if the coroutine is already canceled, it throws `CancellationException`.

One way to make your coroutine cancelable is to check whether the coroutine job is active (still running or completing) or not by using `isActive`. The value of `isActive` will become false once the coroutine job changes its state to `Cancelling`, `Cancelled`, or `Completed`. You can make your coroutine cancelable with `isActive` with the following approaches:

- Perform tasks while `isActive` is true.

- Perform tasks only if `isActive` is true.

- Return or throw an exception if `isActive` is false.

Another function you can also use is `Job.ensureActive()`. It will check whether the coroutine job is active, and if it's not, it will throw `CancellationException`.

Here's an example of how you can make your coroutine cancelable with `isActive`:

```
class SensorActivity : AppCompatActivity() {
    private val scope = CoroutineScope(Dispatchers.IO)
    private lateinit var job: Job

    ...
    private fun processSensorData() {
        job = scope.launch {
            if (isActive) {
                val data = fetchSensorData()
                saveData(data)
```

```
            }
        }
    }

    fun stopProcessingData() {
        job.cancel()
    }


    ...

}
```

The coroutine in the `processSensorData` function will check whether the job is active and will only proceed with the task if the value of `isActive` is true.

Another way to make your coroutine code cancelable is to use suspending functions from the `kotlinx.coroutines` package, such as `yield` or `delay`. The `yield` function yields a thread (or a thread pool) of the current coroutine dispatcher to other coroutines to run.

The `yield` and `delay` functions already check for cancelation and stop the execution or throw `CancellationException`. Thus, you no longer need to manually check for cancelation when you are using them in your coroutines. Here's an example using the preceding code snippet, which has been updated with suspending function delay to make the coroutine cancelable:

```
class SensorActivity : AppCompatActivity() {
    private val scope = CoroutineScope(Dispatchers.IO)
    private lateinit var job: Job

    override fun onCreate(savedInstanceState: Bundle?) {
        ...


        processSensorData()
    }

    private fun processSensorData() {
        job = scope.launch {
            delay(1_000L)
            val data = fetchSensorData()
            saveData(data)
        }
```

```
    }

    fun stopProcessingData() {
        job.cancel()
    }


    ...

}
```

The `delay` suspending function will check whether the coroutine job is canceled and will throw `CancellationException` if it is, making your coroutine cancelable.

Let's learn how to implement a coroutine cancelation for an Android project in the next section.

Exercise 3.01 – canceling coroutines in an Android app

In this exercise, you will work on an application that uses a coroutine that slowly counts down from 100 to 0 and displays the value on `TextView`. You will then add a button to cancel the coroutine to stop the countdown before it reaches 0:

1. Create a new project in Android Studio. Don't change the suggested name of `MainActivity` for the activity.

2. Open the `app/build.gradle` file and add the dependency for `kotlinx-coroutines-android`:

   ```
   implementation 'org.jetbrains.kotlinx:kotlinx-
       coroutines-android:1.6.0'
   ```

   This will add the `kotlinx-coroutines-core` and `kotlinx-coroutines-android` libraries to your project, allowing you to use coroutines in your code.

3. Open the `activity_main.xml` layout file and add an `id` attribute to `TextView`:

   ```
   <TextView
       android:id="@+id/textView"
       style="@style/TextAppearance.AppCompat.Large"
       android:layout_width="wrap_content"
       android:layout_height="wrap_content"
       app:layout_constraintBottom_toBottomOf="parent"
       app:layout_constraintLeft_toLeftOf="parent"
       app:layout_constraintRight_toRightOf="parent"
   ```

```
            app:layout_constraintTop_toTopOf="parent"
            tools:text="0" />
```

The id attribute will allow you to change the content of this TextView later.

4. Open the MainActivity file. Add the following properties to the MainActivity class:

```
private val scope = CoroutineScope(Dispatchers.Main)
private?var job: Job? = null
private lateinit var textView: TextView
private var count = 100
```

5. The first line specifies the scope for the coroutine, CoroutineScope, with Dispatchers. Main as the dispatcher. The second line creates a job property for the coroutine job. The textView property will be used to display the countdown text and count initializes the countdown to 100. In the onCreate function of the MainActivity file, initialize the value for TextView:

```
textView = findViewById(R.id.textView)
```

You will update this textView with the decreasing value of value later.

6. Create a countdown function that will do the counting down of the value:

```
private fun countdown() {
    count--
    textView.text = count.toString()
}
```

This decreases the value of count by 1 and displays it on the text view.

7. In the onCreate function, below the textView initialization, add the following to start the coroutine to count down the value and display it on the text view:

```
job = scope.launch {
    while (count > 0) {
        delay(100)
        countdown()
    }
}
```

This will call the countdown function every 0.1 seconds, which will count down and display the value on the text view.

8. Run the application. You will see that it slowly counts down and displays the value from 100 to 0, similar to the following:

Figure 3.2 – The app counting down from 100 to 0

9. Open the `strings.xml` file and add a string for the button:

```
<string name="stop">Stop</string>
```

You will use this as the text for the button to stop the countdown.

10. Go to the `activity_main.xml` file again and add a button below `TextView`:

```
<Button
        android:id="@+id/button"
        android:layout_width="wrap_content"
        android:layout_height="wrap_content"
        android:layout_marginTop="16dp"
        android:text="@string/stop"
        app:layout_constraintEnd_toEndOf="parent"
        app:layout_constraintStart_toStartOf="parent"
        app:layout_constraintTop_toBottomOf="@id/
textView" />
```

This will add a `Button` below `TextView`. The button will be used to stop the countdown later.

11. Open `MainActivity` and after the job initialization, create a variable for the button:

```
val button = findViewById<Button>(R.id.button)
```

This button, when tapped, will allow the user to stop the countdown.

12. Below that, add a click listener to the button that cancels the job:

```
button.setOnClickListener {
    job?.cancel()
}
```

When you click the button, it will cancel the coroutine.

13. Run the application again. Tap on the **STOP** button and notice that the counting down stops, as shown in the following figure:

Figure 3.3 – Clicking the STOP button cancels the coroutine

Tapping on the **STOP** button cancels the coroutine with the `job.cancel()` call. This works because the coroutine is using the suspending `delay` function, which checks whether the coroutine is active or not.

In this exercise, you have worked on adding code to cancel a running coroutine in an Android app by tapping on a button.

There may be instances where you want to continue work even if you have canceled the job. To ensure the tasks will be done even if the coroutine is canceled, you can use `withContext(NonCancellable)` on the task.

In this section, you learned how to cancel coroutines and how to make sure your coroutine code is cancelable. You will learn how to handle coroutine timeouts in the next section.

Managing coroutine timeouts

In this section, you will learn about timeouts and how you can cancel your coroutines with a timeout. Setting a fixed amount of time after which to stop asynchronous code that is running longer than expected can help you save resources and immediately notify users of any issues.

When your application is doing a background task, you may want to stop it because it is taking too long. You can manually track the time and cancel the task. Or you can use the `withTimeout` suspending function. With the `withTimeout` function, you can set your timeout in milliseconds or `Duration`. Once this timeout is exceeded, it will throw `TimeOutCancellationException`, a subclass of `CancellationException`. Here's an example of how you can use `withTimeout`:

```
class MovieViewModel: ViewModel() {
    init {
        viewModelScope.launch {
            val job = launch {
                withTimeout(5_000L) {
                    fetchMovies()
                }
            }
            ...
        }
    }
}
```

A timeout of 5,000 milliseconds (5 seconds) has been set for the coroutine. If the `fetchMovies` task takes longer than that, the coroutine will time out and throw `TimeoutCancellationException`.

Another function you can use is `withTimeoutOrNull`. It is similar to the `withTimeout` function, but it will return null if the timeout was exceeded. Here's an example of how you can use `withTimeoutOrNull`:

```
class MovieViewModel: ViewModel() {
    init {
        viewModelScope.launch() {
            val job = async {
                fetchMovies()
            }


            val movies = withTimeoutOrNull(5_000L) {
                job.await()
            }
            . . .
        }
    }
    . . .
}
```

The coroutine will return null if `fetchMovies` times out after 5 seconds, and if not, it will return the list of movies fetched.

As you learned in the previous section, the coroutine must be cancelable so that it will be canceled after the timeout. In the next section, you will learn how to handle the cancelation exception from coroutines.

In this section, you have learned about coroutine timeouts and how you can set an amount of time after which to automatically cancel a coroutine.

Catching exceptions in coroutines

In this section, you will learn about coroutine exceptions and how to handle them in your application. As it is always possible that your coroutines will fail, it is important to learn how to catch exceptions so that you can avoid crashes and notify your users.

To handle exceptions in your coroutines, you can simply use `try-catch`. For example, if you have a coroutine started with a `launch` coroutine builder, you can do the following to handle exceptions:

```
class MovieViewModel: ViewModel() {
```

```
    init {
        viewModelScope.launch() {
            try {
                fetchMovies()
            } catch (exception: Exception) {
                Log.e("MovieViewModel",
                    exception.message.toString())
            }
        }
    }
    ...
}
```

If `fetchMovies` has an exception, `ViewModel` will write the exception message to the logs.

If your coroutine was built using the `async` coroutine builder, the exception will be thrown when you call the `await` function on the `Deferred` object. Your code to handle the exception would look like the following:

```
class MovieViewModel: ViewModel() {
    init {
        viewModelScope.launch() {
            val job = async {
                fetchMovies()
            }
            var movies = emptyList<Movie>()
            try {
                movies = job.await()
            } catch (exception: Exception) {
                Log.e("MovieViewModel",
                    exception.message.toString())
            }
        }
    }
    ...
}
```

If an exception is encountered while the `fetchMovies` call is running, the movies list will be an empty list of movies, and `ViewModel` will write the exception message to the logs.

When a coroutine encounters an exception, it will cancel the job and pass on the exception to its parent. This parent coroutine will be canceled, as well as its children. Exceptions in the child coroutines will not affect the parent and its sibling coroutines if you use `SupervisorJob` as follows:

- Creating the coroutine scope with the suspending `supervisorScope{}` builder

- Using `SupervisorJob` for your coroutine scope: `CoroutineScope(SupervisorJob())`

If the exception of your coroutine is a subclass of `CancellationException`, for example, `TimeoutCancellationException` or a custom one you pass to the `cancel` function, the exception will not be transmitted to the parent.

When handling coroutine exceptions, you can also use a single place to handle these exceptions with `CoroutineExceptionHandler`. `CoroutineExceptionHandler` is a coroutine context element that you can add to your coroutine to handle uncaught exceptions. The following lines of code show how you can use it:

```
class MovieViewModel: ViewModel() {
    private val exceptionHandler =
      CoroutineExceptionHandler { _, exception ->
        Log.e("MovieViewModel",
          exception.message.toString())
    }

    private val scope = CoroutineScope(exceptionHandler)

    ...

}
```

The exceptions from the coroutines started from the scope will be handled by `exceptionHandler`, if it's not handled wherever an error could occur, which will write the exception message to the logs.

Let's try to add code to handle exceptions in your coroutines.

Exercise 3.02 – catching exceptions in your coroutines

In this exercise, you will continue working on the application that displays on TextView a number from 100 and slowly decreases it down to 0. You will be adding code to handle exceptions in the coroutine:

1. Open the countdown app you built in the previous exercise.

2. Go the MainActivity file and at the end of the countdown function, add the following to simulate an exception:

    ```
    if ((0..9).random() == 0) throw Exception("An error
        occurred")
    ```

    This will generate a random number from 0 to 9 and if it's 0, it will throw an exception. It will simulate the coroutine encountering an exception.

3. Run the application. It will start to count down and some point later, it will throw the exception and crash the app.

4. Surround the code in your coroutine with a try-catch block to catch the exception in the app:

    ```
    job = scope.launch {
        try {
            while (count > 0) {
                delay(100)
                countdown()
            }
        } catch (exception: Exception) {
            //TODO
        }
    }
    ```

    This will catch the exception from the countdown function. The app will no longer crash but you will need to inform the user about the exception.

5. Inside the catch block, replace //TODO with Snackbar to display the exception message:

    ```
    Snackbar.make(textView, exception.message.toString(),
        Snackbar.LENGTH_LONG).show()
    ```

    This will display a snackbar message with the text An error occurred, which is the message of the exception.

6. Run the application again. It will start to count down but instead of crashing, a snackbar message will be displayed, as shown in the following figure:

Figure 3.4 – Snackbar displayed when the coroutine has encountered the exception

In this exercise, you updated your application so that it can handle exceptions in the coroutines instead of crashing.

In this section, you have learned about coroutine exceptions and how you can catch them in your Android apps.

Summary

In this chapter, you learned about coroutine cancelations. You can cancel coroutines by using the `cancel` or `cancelAndJoin` function from the coroutine job or the `cancel` function from the coroutine scope.

You learned that a coroutine cancelation needs to be cooperative. You also learned how you can change your code to make your coroutine cancelable by using `isActive` checks or by using suspending functions from the `kotlinx.coroutines` package.

Then, you learned about coroutine timeouts. You can set a timeout (in milliseconds or `Duration`) using `withTimeout` or `withTimeoutOrNull`.

You also learned about coroutine exceptions and how to catch them. `try-catch` blocks can be used to handle exceptions. You can also use `CoroutineExceptionHandler` in your coroutine scope to catch and handle exceptions in a single location.

Finally, you worked on an exercise to add cancelation to a coroutine and another exercise to update your code to handle coroutine exceptions.

In the next chapter, you will dive into creating and running tests for the coroutines in your Android projects.

4
Testing Kotlin Coroutines

In the previous chapter, you learned about coroutine cancelations and how you can make your coroutines cancelable. You then learned about coroutine timeouts in milliseconds or **Duration**. Finally, you learned about exceptions and how you can handle them using `try-catch` and `CoroutineExceptionHandler`.

Creating tests is an important part of app development. The more code you write, the higher the chance that there will be bugs and errors. With tests, you can ensure your application works as you have programmed it. You can quickly discover issues and fix them immediately. Tests can make development easier, saving you time and resources. They can also help you refactor and maintain your code with confidence.

In this chapter, you will learn how to test Kotlin coroutines in Android. First, we will update the Android project for testing. We will then proceed with learning the steps to create tests for Kotlin coroutines.

In this chapter, we are going to cover the following topics:

- Setting up an Android project for testing coroutines
- Unit testing suspending functions
- Testing coroutines

By the end of this chapter, you will understand coroutine testing. You will be able to write and run unit and integration tests for the coroutines in your Android applications.

Technical requirements

You will need to download and install the latest version of Android Studio. You can find the latest version at `https://developer.android.com/studio`. For an optimal learning experience, a computer with the following specifications is recommended:

- Intel Core i5 or equivalent or higher
- 4 GB RAM minimum
- 4 GB available space

The code examples for this chapter can be found on GitHub at `https://github.com/PacktPublishing/Simplifying-Android-Development-with-Coroutines-and-Flows/tree/main/Chapter04`.

Setting up an Android project for testing coroutines

In this section, we will start by looking at how you can update your Android app to make it ready for adding and running tests. Once your project is properly set up, it will be easy to add unit and integration tests for your coroutines.

When creating unit tests on Android, you must have the **JUnit 4** testing framework in your project. JUnit is a unit testing framework for Java. It should be automatically included in the `app/build.gradle` dependencies when creating a new Android project in Android Studio.

If your Android project does not have JUnit 4 yet, you can add it by including the following to your `app/build.gradle` dependencies:

```
dependencies {
    ...
    testImplementation 'junit:junit:4.13.2'
}
```

This allows you to use the JUnit 4 framework for your unit tests.

To create mock objects for your tests, you can also use mocking libraries. Mockito is the most popular Java mocking library, and you can use it on Android. To add Mockito to your tests, add the following to the dependencies in your `app/build.gradle` file:

```
dependencies {
    ...
    testImplementation 'org.mockito:mockito-core:4.0.0'
}
```

Adding this dependency allows you to use Mockito to create mock objects for your unit tests in your project.

If you prefer to use Mockito with idiomatic Kotlin code, you can use Mockito-Kotlin. Mockito-Kotlin is a Mockito library that contains helper functions to make your code more Kotlin-like.

To use Mockito-Kotlin in your Android unit tests, you can add the following dependency to your `app/build.gradle` file dependencies:

```
dependencies {
    ...
```

```
        testImplementation 'org.mockito.kotlin:mockito-
            kotlin:4.0.0'
}
```

This will enable you to use Mockito to create mock objects for your tests, using idiomatic Kotlin code.

If you are using both Mockito (`mockito-core`) and Mockito-Kotlin in your project, you can just add the dependency for Mockito-Kotlin. It already has a dependency to `mockito-core`, which it will automatically import.

To test Jetpack components such as `LiveData`, add the `androidx.arch.core:core-testing` dependency:

```
dependencies {
    ...
    testImplementation 'androidx.arch.core:core-
        testing:2.1.0'
}
```

This dependency contains support for testing Jetpack architecture components. It includes JUnit rules such as `InstantTaskExecutorRule` that you can use to test the `LiveData` objects in your code.

Testing coroutines is a bit more complicated than the usual testing. This is because coroutines are asynchronous, tasks can run in parallel, and tasks can take a while before finishing. Your tests must be fast and consistent.

To help you with testing coroutines, you can use the coroutine testing library from the `kotlinx-coroutines-test` package. It contains utility classes to make testing coroutines easier and more efficient. To use it in your Android project, you must add the following to the dependencies in your `app/build.gradle` file:

```
dependencies {
    ...
    testImplementation 'org.jetbrains.kotlinx:kotlinx-
        coroutines-test:1.6.0'
}
```

This will import the `kotlinx-coroutines-test` dependency into your Android project. You will then be able to use the utility classes from the Kotlin coroutine testing library to create unit tests for your coroutines.

If you want to use `kotlinx-coroutines-test` in your Android instrumented tests that will run on an emulator or physical device, you should add the following to your `app/build.gradle` file dependencies:

```
dependencies {
    ...
    androidTestImplementation
        'org.jetbrains.kotlinx:kotlinx-coroutines-test:1.6.0'
}
```

Adding this to your dependencies will allow you to use `kotlinx-coroutines-test` in your instrumented tests.

As of version 1.6.0, the coroutine testing library is still labeled as experimental. You may have to annotate the test classes with the `@ExperimentalCoroutinesApi` annotation, as shown in the following example:

```
@ExperimentalCoroutinesApi
class MovieRepositoryUnitTest {
    ...
}
```

In this section, you learned how to set up your Android project to add tests. You will learn how to create unit tests for suspending functions in the next section.

Unit testing suspending functions

In this section, we will focus on how you can unit test your suspending functions. You can create unit tests for classes such as `ViewModel` that launch a coroutine or have suspending functions.

Creating a unit test for a suspending function is more difficult to write as a suspending function can only be called from a coroutine or another coroutine. What you can do is use the `runBlocking` coroutine builder and call the suspending function from there. For example, say you have a `MovieRepository` class like the following:

```
class MovieRepository (private val movieService:
  MovieService) {
    ...
    private val movieLiveData =
      MutableLiveData<List<Movie>>()

    fun fetchMovies() {
```

```
        . . .
        val movies = movieService.getMovies()
        movieLiveData.postValue(movies.results)
    }
}
```

This `MovieRepository` has a suspending function called `fetchMovies`. This function gets the list of movies by calling the `getMovies` suspending function from `movieService`.

To create a test for the `fetchMovies` function, you can use `runBlocking` to call the suspending function, like the following:

```
class MovieRepositoryTest {
    . . .

    @Test
    fun fetchMovies() {
        . . .
        runBlocking {
            . . .
            val movieLiveData =
              movieRepository.fetchMovies()
            assertEquals(movieLiveData.value, movies)
        }
    }
}
```

Using the `runBlocking` coroutine builder allows you to call suspending functions and do the assertion checks.

The `runBlocking` coroutine builder is useful for testing. However, there are times when it can be slow because of delays in the code. Your unit tests must ideally be able to run as fast as possible. The coroutine testing library can help you with its `runTest` coroutine builder. It is the same as the `runBlocking` coroutine builder except it runs the suspending function immediately and without delays.

Replacing `runBlocking` with `runTest` in the previous example would make your test look like the following:

```
@ExperimentalCoroutinesApi
class MovieRepositoryTest {
```

```
...


@Test
fun fetchMovies() {

    ...
    runTest {

        ...
        val movieLiveData =
          movieRepository.fetchMovies()
        assertEquals(movieLiveData.value, movies)

    }

}

}
```

The runTest function allows you to call the movieRepository.fetchMovies() suspending function and then check the result of the operation.

In this section, you learned about writing unit tests for suspending functions in your Android project. In the next section, you will learn about testing coroutines.

Testing coroutines

In this section, we will focus on how you can test your coroutines. You can create tests for classes such as ViewModel that launch a coroutine.

For coroutines launched using Dispatchers.Main, your unit tests will fail with the following error message:

```
java.lang.IllegalStateException: Module with the Main
dispatcher had failed to initialize. For tests Dispatchers.
setMain from kotlinx-coroutines-test module can be used
```

This exception happens because Dispatchers.Main uses Looper.getMainLooper(), the application's main thread. This main looper is not available in Android for local unit tests. To make your tests work, you must use the Dispatchers.setMain extension function to change the Main dispatcher. For example, you can create a function in your test class that will run before your tests:

```
@Before
fun setUp() {
```

```
        Dispatchers.setMain(UnconfinedTestDispatcher())
}
```

The `setUp` function will run before the tests. It will change the main dispatcher to another dispatcher for your test.

`Dispatchers.setMain` will change all subsequent uses of `Dispatchers.Main`. After the test, you must change the `Main` dispatcher back with a call to `Dispatchers.resetMain()`. You can do something like the following:

```
@After
fun tearDown() {
    Dispatchers.resetMain()
}
```

After the tests have run, the `tearDown` function will be called, which will reset the `Main` dispatcher.

If you have many coroutines to test, copying and pasting this boilerplate code in each test class is not ideal. You can make a custom JUnit rule instead that you can reuse in your test classes. This JUnit rule must be in the root folder of your test source set, as shown in *Figure 4.01*:

Figure 4.1 – Custom TestCoroutineRule in the root test folder

An example of a custom JUnit rule that you can write to reuse for automatically setting `Dispatchers.setMain` and `Dispatchers.resetMain` is this `TestCoroutineRule`:

```
@ExperimentalCoroutinesApi
class TestCoroutineRule(val dispatcher: TestDispatcher =
  UnconfinedTestDispatcher()):
  TestWatcher() {
  override fun starting(description: Description?) {
      super.starting(description)
      Dispatchers.setMain(dispatcher)
  }

  override fun finished(description: Description?) {
      super.finished(description)
      Dispatchers.resetMain()
  }
}
```

This custom JUnit rule will allow your test to automatically call `Dispatchers.setMain` before the tests and `Dispatchers.resetMain` after the tests.

You can then use this `TestCoroutineRule` in your test classes by adding the `@get:Rule` annotation:

```
@ExperimentalCoroutinesApi
class MovieRepositoryTest {
    @get:Rule
    var coroutineRule = TestCoroutineRule()

    ...
}
```

With this code, you will not need to add the `Dispatchers.setMain` and `Dispatchers.resetMain` function calls every time in your test classes.

When testing your coroutines, you must replace your coroutine dispatchers with a `TestDispatcher` for testing. To be able to replace your dispatchers, your code should have a way to change the dispatcher that will be used for the coroutines. For example, this `MovieViewModel` class has a property for setting the dispatcher:

```
class MovieViewModel(private val dispatcher:
    CoroutineDispatcher = Dispatchers.IO): ViewModel() {

    ...


    fun fetchMovies() {
        viewModelScope.launch(dispatcher) {
            ...
        }
    }
}
```

`MovieViewModel` uses the dispatcher specified in its constructor or the default value (`Dispatchers.IO`) for launching the coroutine.

In your test, you can then set a different `Dispatcher` for testing purposes. For the preceding `ViewModel`, your test could initialize `ViewModel` with a different dispatcher, as shown in the following example:

```
@ExperimentalCoroutinesApi
class MovieViewModelTest {
    ...


    @Test
    fun fetchMovies() {
        ...
        runTest {
            ...
            val viewModel =
                MovieViewModel(UnconfinedTestDispatcher())
            viewModel.fetchMovies()
            ...
        }
    }
}
```

The `viewModel` in `MovieViewModelTest`'s `fetchMovies` test was initialized with `UnconfinedTestDispatcher` as the coroutine dispatcher for testing purposes.

In the previous examples, you used `UnconfinedTestDispatcher` as the `TestDispatcher` for the test. There are two available implementations of `TestDispatcher` in the `kotlinx-coroutines-test` library:

- `StandardTestDispatcher`: Does not run coroutines automatically, giving you full control over execution order

- `UnconfinedTestDispatcher`: Runs coroutines automatically; offers no control over the order in which the coroutines will be launched

Both `StandardTestDispatcher` and `UnconfinedTestDispatcher` have constructor properties: `scheduler` for `TestCoroutineScheduler` and name for identifying the dispatcher. If you do not specify the scheduler, `TestDispatcher` will create a `TestCoroutineScheduler` by default.

The `TestCoroutineScheduler` of the `StandardTestDispatcher` controls the execution of the coroutine. `TestCoroutineScheduler` has three functions you can call to control the execution of the tasks:

- `runCurrent()`: Runs the tasks that are scheduled until the current virtual time

- `advanceUntilIdle()`: Runs all pending tasks

- `advanceTimeBy(milliseconds)`: Runs pending tasks until current virtual advances by the specified milliseconds

`TestCoroutineScheduler` also has a `currentTime` property that specifies the current virtual time in milliseconds. When you call functions such as `advanceTimeBy`, it will update the `currentTime` property of the scheduler.

The `runTest` coroutine builder creates a coroutine with a coroutine scope of `TestScope`. This `TestScope` has a `TestCoroutineScheduler` (`testScheduler`) that you can use to control the execution of tasks.

This `testScheduler` also has extension property called `currentTime` and the `runCurrent`, `advanceUntilIdle`, and `advanceTimeBy` extension functions, which simplifies calling these functions from the `testScheduler` of the `TestScope`.

Using `runTest` with a `TestDispatcher` allows you to test cases when there are time delays in the coroutine and you want to test a line of code before moving on to the next ones. For example, if your `ViewModel` has a `loading` Boolean variable that is set to `true` before a network operation and then is reset to `false` afterward, your test for the `loading` variable could look like this:

```
@Test
fun loading() {
    val dispatcher = StandardTestDispatcher()

    runTest() {
        val viewModel = MovieViewModel(dispatcher)

        viewModel.fetchMovies()
        dispatcher.scheduler.advanceUntilIdle()
        assertEquals(false, viewModel.loading.value)
    }
}
```

This test uses `StandardTestDispatcher` so you can control the execution of the tasks. After calling `fetchMovies`, you call `advanceUntilIdle` on the dispatcher's `scheduler` to run the task, which will set the `loading` value to `false` after completion.

In this section, you learned about adding tests for your coroutines. Let's test what we have learned so far by adding some tests to existing coroutines in an Android project.

Exercise 4.01 – adding tests to coroutines in an Android app

For this exercise, you will be continuing the movie app that you worked on in *Exercise 2.01, Using coroutines in an Android app*. This application displays the movies that are currently playing in cinemas. You will be adding unit tests for the coroutines in the project by following these steps:

1. Open the movie app you worked on in *Exercise 2.01, Using coroutines in an Android app*, in Android Studio.

2. Go to the `app/build.gradle` file and add the following dependencies, which will be used for the unit test:

```
testImplementation 'org.mockito.kotlin:mockito-
    kotlin:4.0.0'
testImplementation 'androidx.arch.core:core-
```

```
testing:2.1.0'
testImplementation 'org.jetbrains.kotlinx:kotlinx-
   coroutines-test:1.6.0'
```

The first line will add Mockito-Core and Mockito-Kotlin, the second line will add the architecture testing library, and the last line will add the Kotlin coroutine testing library. You will be using these for the unit tests you will add to the Android project.

3. In `app/src/test/resources`, create a `mockito-extensions` directory. In that directory, create a new file named `org.mockito.plugins.MockMaker`, as shown in *Figure 4.2*:

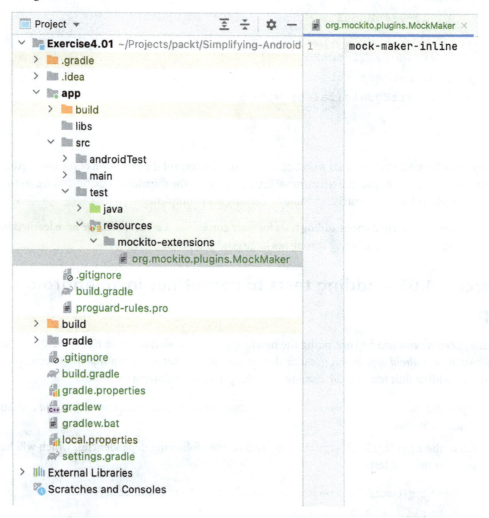

Figure 4.2 – The file you need to add to the app/src/test/mockito-extensions directory

4. In the `app/src/test/mockito-extensions/org.mockito.plugins.MockMaker` file, add the following content:

   ```
   mock-maker-inline
   ```

   This will allow you to create mocks using Mockito for final classes in your code. Without this, your test will fail with the following error message:

   ```
   Mockito cannot mock/spy because : final class
   ```

5. You will first add a unit test for the `MovieRepository` class. In `app/src/test`, create a test class called `MovieRepositoryTest` and add the `@OptIn(ExperimentalCoroutinesApi::class)` annotation to this class:

   ```
   @OptIn(ExperimentalCoroutinesApi::class)
   class MovieRepositoryTest {

       ...

   }
   ```

   This will be the test class for `MovieRepository`. The `ExperimentalCoroutinesApi` `OptIn` annotation was added as some of the classes in the `kotlinx-coroutines-test` library are still marked as experimental.

6. Inside the `MovieRepositoryTest` class, add a JUnit test rule for `InstantTaskExecutorRule`:

   ```
   @get:Rule
   val rule = InstantTaskExecutorRule()
   ```

   `InstantTaskExecutorRule` allows the test to execute the tasks synchronously. This is needed for the `LiveData` objects in `MovieRepository`.

7. Create a test function called `fetchMovies` to test the `fetchMovies` suspending function from `MovieRepository`, successfully retrieving a list of movies:

   ```
   @Test
   fun fetchMovies() {

       ...

   }
   ```

   This will be the first test for `MovieRepository.fetchMovies`: a success scenario that displays a list of movies.

8. In the MovieRepositoryTest class' fetchMovies function, add the following code to mock MovieRepository and MovieService:

```
@Test
fun fetchMovies() {
    val movies = listOf(Movie(id = 3), Movie(id = 4))
    val response = MoviesResponse(1, movies)

    val movieService: MovieService = mock {
        onBlocking { getMovies(anyString()) } doReturn
            response
    }
    val movieRepository =
        MovieRepository(movieService)
}
```

This will mock MovieService so that when its getMovies function is called, it will always return the movies list we provided.

9. At the end of the fetchMovies function of MovieRepositoryTest, add the following to test that calling fetchMovies from the MovieRepository class returns the list of movies we expect it to return:

```
@Test
fun fetchMovies() {

    ...

    runTest {
        movieRepository.fetchMovies()
        val movieLiveData = movieRepository.movies
        assertEquals(movies, movieLiveData.value)
    }
}
```

This will call the fetchMovies function from the MovieRepository class, which will call getMovies from MovieService. We are checking whether it indeed returns the list of movies that we set in the mocked MovieService earlier.

10. Run the MovieRepositoryTest class. MovieRepositoryTest should pass and there should be no errors.

11. Create another test function called `fetchMoviesWithError` in the `MovieRepositoryTest` class to test the `fetchMovies` suspending function from the `MovieRepository` failing to retrieve a list of movies:

```
@Test
fun fetchMoviesWithError() {
    ...
}
```

This will test the case when `MovieRepository` fails while retrieving the list of movies.

12. In the `MovieRepositoryTest` class' `fetchMoviesWithError` function, add the following:

```
@Test
fun fetchMoviesWithError() {
    val exception = "Test Exception"
    val movieService: MovieService = mock {
        onBlocking { getMovies(anyString()) } doThrow
            RuntimeException(exception)
    }
    val movieRepository =
        MovieRepository(movieService)
}
```

This will mock `MovieService` so that when its `getMovies` function is called, it will always throw an exception with the message `Test Exception`.

13. At the end of the `fetchMoviesWithError` function of `MovieRepositoryTest`, add the following to test that calling `fetchMovies` from the `MovieRepository` class returns the list of movies we expect it to return:

```
@Test
fun fetchMovies() {
    ...

    runTest {
        movieRepository.fetchMovies()

        val movieLiveData = movieRepository.movies
        assertNull(movieLiveData.value)
```

```
          val errorLiveData = movieRepository.error
          assertNotNull(errorLiveData.value)
          assertTrue(errorLiveData.value.toString()
            .contains(exception))
          }
    }
```

This will call the `fetchMovies` function from the `MovieRepository` class, which will call the `getMovies` from the `MovieService` that will always throw an exception when called.

In the first assertion, we are checking that `movieLiveData` is null as there were no movies fetched. The second assertion checks that `errorLiveData` is not null as there was an exception. The last assertion checks that `errorLiveData` contains the `Test Exception` message we set in the previous step.

14. Run the `MovieRepositoryTest` test. Both the `fetchMovies` and `fetchMoviesWithError` tests should have no errors and both should pass.

15. We will then create a test for `MovieViewModel`. First, we would need to update `MovieViewModel` so that we can change the dispatcher that the coroutine runs on. Open the `MovieViewModel` class and update its constructor by adding a dispatcher property to set the coroutine dispatcher:

```
class MovieViewModel(private val movieRepository:
  MovieRepository, private val dispatcher:
  CoroutineDispatcher = Dispatchers.IO) : ViewModel()
  {
    ...
  }
```

This will allow you to change the dispatcher of `MovieViewModel` with another dispatcher, which you will be doing in the tests.

16. In the `fetchMovies` function, change the `launch` coroutine builder to use the `dispatcher` from the constructor instead of the hardcoded dispatcher:

```
viewModelScope.launch(dispatcher) {
    ...
  }
```

This updates the code to use the `dispatcher` set from the constructor or the default dispatcher (`Dispatchers.IO`). You can now create a unit test for the `MovieViewModel` class.

17. In the `app/src/test` directory, create a test class named `MovieViewModelTest` for `MovieViewModel` and add the `@OptIn(ExperimentalCoroutinesApi::class)` annotation to the class:

```
@OptIn(ExperimentalCoroutinesApi::class)
class MovieViewModelTest {

    ...

}
```

This will be the test class for `MovieViewModel`. The `ExperimentalCoroutinesApi` annotation was added as some of the classes in the `kotlinx-coroutines-test` library are still experimental.

18. Inside the `MovieViewModelTest` class, add a JUnit test rule for `InstantTaskExecutorRule`:

```
@get:Rule
val rule = InstantTaskExecutorRule()
```

The `InstantTaskExecutorRule` in the unit test executes the tasks synchronously. This is for the `LiveData` objects in `MovieViewModel`.

19. Create a test function called `fetchMovies` to test the `fetchMovies` suspending function from `MovieViewModel`:

```
@Test
fun fetchMovies() {
    val expectedMovies =
        MutableLiveData<List<Movie>>()
    expectedMovies.postValue(listOf(Movie
        (title = "Movie")))

    val movieRepository: MovieRepository = mock {
        onBlocking { movies } doReturn expectedMovies
    }
}
```

This will mock `MovieRepository` so that its `movies` property will always return the `expectedMovies` as its value.

20. At the end of the `fetchMovies` test of `MovieViewModelTest`, add the following to test that `MovieViewModel`'s movies will be equal to `expectedMovies`:

```
@Test
fun fetchMovies() {

    . . .


    val movieViewModel =
        MovieViewModel(movieRepository)
    assertEquals(expectedMovies.value,
        movieViewModel.movies.value)
}
```

This creates a `MovieViewModel` using the mocked `MovieRepository`. We are checking that the value of `MovieViewModel`'s `movies` is equal to the `expectedMovies` value we set to the mocked `MovieRepository`.

21. Run `MovieViewModelTest` or all the tests (`MovieRepositoryTest` and `MovieViewModelTest`). All tests should pass.

22. Create another test function called `loading` in `MovieViewModelTest` to test the `loading` LiveData in `MovieViewModel`:

```
@Test
fun loading() {

    . . .

}
```

This will test the `loading` LiveData property of `MovieViewModel`. The loading property is `true` while fetching the movies and displays the `ProgressBar`. It becomes `false` and hides the `ProgressBar` after successfully fetching the movies or when an error is encountered.

23. In the `loading` test function, add the following to mock `MovieRepository` and initialize a dispatcher that will be used for `MovieViewModel`:

```
@Test
fun loading() {
    val movieRepository: MovieRepository = mock()
    val dispatcher = StandardTestDispatcher()


    . . .

}
```

This will mock `MovieRepository` and create a dispatcher of the `StandardTestDispatcher` type that will be used for the `MovieViewModel` test. This will allow you to control the execution of the task, which will be used later to check the value of `MovieViewModel`'s `loading` property.

24. At the end of the `loading` test function, add the following to test the loading `MovieViewModel`'s `loading` property:

```
@Test
fun loading() {

    ...

    runTest {
        val movieViewModel =
            MovieViewModel(movieRepository, dispatcher)

        movieViewModel.fetchMovies()
        assertTrue( movieViewModel.loading.value ==
            true)
        dispatcher.scheduler.advanceUntilIdle()
        assertFalse(movieViewModel.loading.value ==
            true)
    }
}
```

This will create a `MovieViewModel` with the mock `MovieRepository` and `dispatcher` you created in the previous step. Then, `fetchMovies` will be called from `MovieViewModel` to fetch the list of movies.

The first assertion checks whether the `loading` value is `true`. We then used `advanceUntilIdle` from the dispatcher's `scheduler` to execute all the tasks. This should change the `loading` value to `false`. The last line checks this indeed happens.

25. Run both `MovieRepositoryTest` and `MovieViewModelTest`. All the tests should pass.

In this exercise, you worked on an Android project that uses coroutines and you have added unit tests for these coroutines.

Summary

This chapter focused on testing coroutines in your Android app. You started with learning how to set up your Android project in preparation for adding tests for your coroutines. The coroutines testing library (`kotlinx-coroutines-test`) helps you to create tests for your coroutines.

You learned how to add unit tests for your suspending functions. You can use `runBlocking` and `runTest` to test code that calls suspending functions. `runTest` runs the code immediately, without delays.

Then, you learned how to test coroutines. You can change the dispatcher in your test with a `TestDispatcher` (`StandardTestDispatcher` or `UnconfinedTestDispatcher`). `TestCoroutineScheduler` allows you to control the execution of the coroutine task.

Finally, you worked on an exercise where you added unit tests for coroutines in an existing Android project.

In the next chapter, you will explore Kotlin Flows and learn how you can use them for asynchronous programming in Android.

Further reading

This book assumes that you already have knowledge of testing Android applications. If you would like to learn more about Android testing, you can read *Chapter 9, Unit Tests and Integration Tests with JUnit, Mockito, and Espresso,* from the book *How to Build Android Apps with Kotlin (Packt Publishing, 2021, ISBN 9781838984113).* You can also check the Android testing documentation at `https://developer.android.com/training/testing`.

As of the time of writing, the coroutine testing library is still marked as experimental. Before the library becomes stable later, there might be some code-breaking changes to the classes. You can check the latest version of the library on GitHub at `https://github.com/Kotlin/kotlinx.coroutines/tree/master/kotlinx-coroutines-test` to find the latest information about the coroutine testing library.

Part 2 –
Kotlin Flows
on Android

In this part, we will learn how to fetch data using Kotlin Flows. We will also discuss how to combine flows, handle cancelations and exceptions, and create tests for them.

This section comprises the following chapters:

5
Using Kotlin Flows

In the previous three chapters, we dove into Kotlin coroutines and learned how we can use them for asynchronous programming in Android. We learned about coroutine builders, scopes, dispatchers, contexts, and jobs. We then learned how to handle coroutine cancelations, timeouts, and exceptions. We also learned how to create tests for coroutines in your code.

In the next three chapters, we will focus on Kotlin Flow, a new asynchronous stream library built on top of Kotlin coroutines. A flow can emit multiple values over a length of time instead of just a single value. You can use Flows for streams of data, such as real-time location, sensor readings, and live database values.

In this chapter, we will explore Kotlin Flows. We will start by building Kotlin Flows. Then, we will look into the various operators you can use for transforming, combining, buffering, and doing more with Flows. Finally, we will learn about StateFlows and SharedFlows.

This chapter covers the following main topics:

- Using Flows in Android
- Creating Flows with Flow builders
- Using operators with Flows
- Buffering and combining Flows
- Exploring StateFlow and SharedFlow

By the end of this chapter, you will have a deeper understanding of using Kotlin Flows. You will be able to use Flows for various cases in your Android apps. You will also learn about flow builders, operators, combining flows, StateFlow, and SharedFlow.

Technical requirements

You will need to download and install the latest version of Android Studio. You can find the latest version at `https://developer.android.com/studio`. For an optimal learning experience, a computer with the following specifications is recommended:

- Intel Core i5 or equivalent or higher

- 4 GB RAM minimum

- 4 GB available space

The code examples for this chapter can be found on GitHub at `https://github.com/PacktPublishing/Simplifying-Android-Development-with-Coroutines-and-Flows/tree/main/Chapter05`.

Using Flows in Android

In this section, we will start by using flows in Android for asynchronous programming. Flows are ideal for the parts of your application that involve live data updates.

A Flow of data is represented by the **kotlinx.coroutines.flow.Flow** interface. Flows emit multiple values of the same type one at a time. For example, `Flow<String>` is a flow that emits string values.

Android Jetpack libraries such as Room, Paging, DataStore, WorkManager, and Jetpack Compose include built-in support for Flow.

The Room database library added support for Flows, starting with version 2.2. This allows you to be notified of changes in the database values by using Flows.

If your Android application uses a **Data Access Object** (**DAO**) to display a list of movies, your project can have a DAO such as the following:

```
@Dao
interface MovieDao {

    @Query("SELECT * FROM movies")
    fun getMovies(): List<Movie>

    ...

}
```

By calling the `getMovies` function from `MovieDao`, you can get the list of movies from the database.

The preceding code will only fetch the list of movies once, after calling `getMovies`. You may want your application to automatically update the list of movies whenever a movie in the database has been added, removed, or updated. You can do that by using Room-KTX and changing your `MovieDao` to use Flow for `getMovies`:

```
@Dao
interface MovieDao {

    @Query("SELECT * FROM movies")
    fun getMovies(): Flow<List<Movie>>


    ...

}
```

With this code, every time the `movies` table has a change, `getMovies` will emit a new list containing the list of movies from the database. Your application can then use that to automatically update the movies displayed in your List or `RecyclerView`.

If you are using `LiveData` and want to convert `LiveData` to `Flow`, or `Flow` to `LiveData`, you can use the LiveData KTX.

To convert `LiveData` to `Flow`, you can use the `LiveData.asFlow()` extension function. With the `Flow.asLiveData()` extension function to convert `Flow` to `LiveData`. You can add LiveData KTX to your project by including the following to your `app/build.gradle` dependencies:

```
dependencies {
    ...


    implementation 'androidx.lifecycle:lifecycle-livedata-
        ktx:2.2.0'
}
```

This adds the LiveData KTX to your project, allowing you to use the `asFlow()` and `asLiveData()` extension functions to convert `LiveData` to `Flow` and `Flow` to `LiveData`.

Third-party Android libraries now also support Flows; some functions can return Flow objects. If you are using RxJava 3 in your project, you can use the **kotlinx-coroutines-rx3** library to convert `Flow` to `Flowable` or `Observable` and vice versa.

A flow will only start emitting values when you call the `collect` function. The `collect` function is a suspending function, so you should call it from a coroutine or another suspending function.

In the following example, the `collect()` function was called from the coroutine created using the `launch` coroutine builder from `lifecycleScope`:

```
class MainActivity : AppCompatActivity() {

    ...


    override fun onCreate(savedInstanceState: Bundle?) {
        ...
        lifecycleScope.launch {
            viewModel.fetchMovies().collect { movie ->
                Log.d("movies", "${movie.title}")
            }
        }
    }
}

class MovieViewModel : ViewModel() {

    ...

    fun fetchMovies(): Flow<Movie> {
        ...
    }
}
```

In this example, the `collect{}` function was called on `Flow<Movie>` and returned by calling `viewModel.fetchMovies()`. This will cause the Flow to start emitting values, and you can then process each value.

The collection of the flow occurs in `CoroutineContext` of the parent coroutine. In the previous example, the coroutine context is from `viewModelScope`.

To change `CoroutineContext` where the Flow is run, you can use the `flowOn()` function. If you want to change `Dispatcher` on the Flow in the previous example to `Dispatchers.IO)`, you can use `flowOn(Dispatchers.IO)`, as shown in the following example:

```
class MainActivity : AppCompatActivity() {

    ...
```

```
    override fun onCreate(savedInstanceState: Bundle?) {
        ...
        lifecycleScope.launch {
            viewModel.fetchMovies()
                .flowOn(Dispatchers.IO)
                .collect { movie ->
                    Log.d("movies", "${movie.title}")
                }
        }
    }
}
```

Here, before collecting the Flow, the dispatcher where the Flow is run was changed to Dispatchers. IO by calling flowOn with Dispatchers.IO.

When you call flowOn, it will only change the preceding functions or operators and not the ones after you called it. In the following example, a map operator was called after the flowOn call to change the dispatcher, so its context won't be changed:

```
class MainActivity : AppCompatActivity() {

    ...

    override fun onCreate(savedInstanceState: Bundle?) {
        ...
        lifecycleScope.launch {
            viewModel.fetchMovies()
                .flowOn(Dispatchers.IO)
                .map { ... }
                .collect { movie ->
                    Log.d("movies", "${movie.title}")
                }
        }
    }
}
```

In this example, flowOn will only change the context of the ones preceding the call, so the map call will not be changed. It will still use the original context (which is the one from lifecycleScope).

In Android, you can collect Flow in the Fragment or Activity classes to display the data in the UI. If the UI goes to the background, your Flow will keep on collecting the data. Your app must not continue collecting the Flow and updating the screen to prevent memory leaks and avoid wasting resources.

To safely collect flows in the Android UI layer, you would need to handle the lifecycle changes yourself. You can also use `Lifecycle.repeatOnLifecycle` and `Flow.flowWithLifecycle`, which are available in the **lifecycle-runtime-ktx** library, starting with version 2.4.0. To add it to your project, you can add the following to your `app/build.gradle` dependencies:

```
dependencies {
    ...
    implementation 'androidx.lifecycle:lifecycle-runtime-
       ktx:2.4.1
}
```

This adds the **lifecycle-runtime-ktx** library to your project, allowing you to use `Lifecycle.repeatOnLifecycle` and `Flow.flowWithLifecycle`.

`Lifecycle.repeatOnLifecycle(state, block)` suspends the parent coroutine until the lifecycle is destroyed and executes the suspending `block` of code when the lifecycle is at least in `state` you set. When the lifecycle moves out of the state, `repeatOnLifecycle` will stop the Flow and restart it when the lifecycle moves back to the said state.

If you used **Lifecycle.State.STARTED** for the state, your `repeatOnLifecycle` will start collecting the Flow whenever the lifecycle is started. It will stop when the lifecycle is stopped, when the `onStop()` of the lifecycle is called.

When you use **Lifecycle.State.RESUMED** for the state, your `repeatOnLifecycle` will start collecting the Flow every time the lifecycle is resumed and will stop when the lifecycle is paused or when `onPause()` is called.

It is recommended to call `Lifecycle.repeatOnLifecycle` on the activity's `onCreate` or on the fragment's `onViewCreated` functions.

The following shows how you can use `Lifecycle.repeatOnLifecycle` in your Android project:

```
class MainActivity : AppCompatActivity() {
    ...

    override fun onCreate(savedInstanceState: Bundle?) {
        ...
        lifecycleScope.launch {
            repeatOnLifecycle(Lifecycle.State.STARTED) {
```

```
                    viewModel.fetchMovies()
                        .collect { movie ->
                            Log.d("movies", "${movie.title}")
                        }
                }
            }
        }
    }
}
```

Here, we used `repeatOnLifecycle` with **Lifecycle.State.STARTED** to start collecting the Flow of movies when the lifecycle is started and stop when the lifecycle is stopped.

You can use `Lifecycle.repeatOnLifecycle` to collect more than one Flow. To do so, you must collect them in parallel in different coroutines:

```
class MainActivity : AppCompatActivity() {
    ...

    override fun onCreate(savedInstanceState: Bundle?) {
        ...
        lifecycleScope.launch {
            repeatOnLifecycle(Lifecycle.State.STARTED) {
                launch {
                    viewModel.fetchMovies().collect { movie ->
                        Log.d("movies", "${movie.title}")
                    }
                }

                launch {
                    viewModel.fetchTVShows.collect { show ->
                        Log.d("tv shows", "${show.title}")
                    }
                }
            }
        }
    }
}
```

Here, there are two Flows: one to collect the movies and the other to collect the TV shows. The collections of the Flow are started from separate `launch` coroutine builders.

If you only have one Flow to collect, you can also use `Flow.flowWithLifecycle`. This emits values from the upstream Flow (the Flow and operators preceding the call) when the lifecycle is at least in **Lifecycle.State.STARTED** or the state you set. It uses `Lifecycle.repeatOnLifecycle` internally. You can use `Flow.flowWithLifecycle` as shown in the following code:

```
class MainActivity : AppCompatActivity() {
    ...

    override fun onCreate(savedInstanceState: Bundle?) {
        ...
        lifecycleScope.launch {
            viewModel.fetchMovies()
                .flowWithLifecycle(lifecycle,
                    Lifecycle.State.STARTED)
                .collect { movie ->
                    Log.d("movies", "${movie.title}")
                }
        }
    }
}
```

In this example, you used `flowWithLifecycle` with **Lifecycle.State.STARTED** to start collecting the Flow of movies when the lifecycle is started and stop if the lifecycle is stopped.

In this section, you have learned about using Kotlin Flows in your Android app. You can use Flow in Android Jetpack libraries such as Room and even in third-party libraries. To safely collect flows in the UI layer and prevent memory leaks and avoid wasting resources, you can use `Lifecycle.repeatOnLifecycle` and `Flow.flowWithLifecycle`.

In the next section, we will be looking into the different Flow builders you can use to create Flows for your application.

Creating Flows with Flow builders

In this section, we will start by looking at creating Flows. To create a Flow, you can use a Flow builder.

The Kotlin Flow API has flow builders that you can use to create Flows. The following are the Kotlin Flow builders you can use:

- `flow {}`
- `flowOf()`
- `asFlow()`

The `flow` builder function creates a new Flow from a suspendable lambda block. Inside the block, you can send values using the `emit` function. For example, this `fetchMovieTitles` function of `MovieViewModel` returns `Flow<String>`:

```
class MovieViewModel : ViewModel() {

    ...


    fun fetchMovieTitles(): Flow<String> = flow {
        val movies = fetchMoviesFromNetwork()
        movies.forEach { movie ->
            emit(movie.title)
        }
    }

    private fun fetchMoviesFromNetwork(): List<Movie> {

        ...

    }

}
```

In this example, `fetchMovieTitles` created a Flow with the movie titles. It iterated over the list of movies from `fetchMoviesFromNetwork` and, for each movie, emitted the movie's title with the `emit` function.

With the `flowOf` function, you can create a Flow that produces the specified value or `vararg` values. In the following example, the `flowOf` function is used to create a Flow of the titles of the top three movies:

```
class MovieViewModel : ViewModel() {

    ...


    fun fetchTop3Titles(): Flow<List<Movie>> {
```

```
        val movies = fetchMoviesFromNetwork().sortedBy {
            it.popularity }
        return flowOf(movies[0].title,
            movies[1].title,
            movies[2].title)
    }


    private fun fetchMoviesFromNetwork(): List<Movie> {
        ...
    }

}
```

Here, `fetchTop3Titles` uses `flowOf` to create a Flow containing the titles of the first three movies.

The `asFlow()` extension function allows you to convert a type into a Flow. You can use this on sequences, arrays, ranges, collections, and functional types. For example, this `MovieViewModel` has `fetchMovieIds` that returns `Flow<Int>`, containing the movie IDs:

```
class MovieViewModel : ViewModel() {

    ...


    private fun fetchMovieIds(): Flow<Int> {
        val movies: List<Movie> = fetchMoviesFromNetwork()
        return movies.map { it.id }.asFlow()
    }

    private fun fetchMoviesFromNetwork(): List<Movie> {
        ...
    }

}
```

In this example, we used a `map` function on the list of movies to create a list of the movie IDs. The list of movie IDs was then converted to `Flow<String>` by using the `asFlow()` extension function on it.

In this section, we learned how you can create Flows with Flow Builders. In the next section, we will check out the various Kotlin Flow operators you can use to transform, combine, and do more with Flows.

Using operators with Flows

In this section, we will focus on the various Flow operators. Kotlin Flow has built-in operators that you can use with Flows. We can collect flows with terminal operators and transform Flows with Intermediate operators.

Collecting Flows with terminal operators

In this section, we will explore the terminal operators you can use on Flows to start the collection of a Flow. The `collect` function we used in the previous examples is the most used terminal operator. However, there are other built-in terminal Flow operators.

The following are the built-in terminal Flow operators you can use to start the collection of the Flow:

- `toList`: Collects the Flow and converts it into a list
- `toSet`: Collects the Flow and converts it into a set
- `toCollection`: Collects the Flow and converts it into a collection
- `count`: Returns the number of elements in the Flow
- `first`: Returns the Flow's first element or throws a **NoSuchElementException** if the Flow was empty
- `firstOrNull`: Returns the Flow's first element or null if the Flow was empty
- `last`: Returns the Flow's last element or throws a **NoSuchElementException** if the Flow was empty
- `lastOrNull`: Returns the Flow's last element or null if the Flow was empty
- `single`: Returns the single element emitted or throws an exception if the Flow was empty or had more than one value
- `singleOrNull`: Returns the single element emitted or null if the Flow was empty or had more than one value
- `reduce`: Applies a function to each item emitted, starting from the first element, and returns the accumulated result
- `fold`: Applies a function to each item emitted, starting from the initial value set, and returns the accumulated result

These terminal Flow operators work like the Kotlin collection functions with the same name in the standard Kotlin library.

In the following example, the `firstOrNull` terminal operator is used instead of the `collect` operator to collect the Flow from `ViewModel`:

```
class MainActivity : AppCompatActivity() {

    ...

    override fun onCreate(savedInstanceState: Bundle?) {
        ...
        lifecycleScope.launch {
            repeatOnLifecycle(Lifecycle.State.STARTED) {
                val topMovie =
                    viewModel.fetchMovies().firstOrNull()
                displayMovie(topMovie)
            }
        }
    }
}
```

Here, `firstOrNull` was used on the Flow to get the first item (or null if the Flow was empty), which represents the top movie. It will then be displayed on the screen.

In this section, you learned about the Flow terminal operators you can use to start collecting from a Flow. In the next section, we will learn how to transform Flows with Intermediate operators.

Transforming Flows with Intermediate operators

In this section, we will focus on Intermediate flow operators that you can use to transform Flows. With Intermediate operators, you can return a new Flow based on the original one.

Intermediate operators allow you to modify a Flow and return a new one. You can chain various operators, and they will be applied sequentially.

You can transform the Flow by applying operators on them, as you can do with Kotlin collections. The following Intermediate operators work similarly to the Kotlin collection functions with the same name:

- `filter`: Returns a Flow that selects only the values from the Flow that meet the condition you passed

- `filterNot`: Returns a Flow that selects only the values from the Flow that do not meet the condition you passed

- `filterNotNull`: Returns a Flow that only includes values from the original Flow that are not null

- `filterIsInstance`: Returns a Flow that only includes values from the Flow that are instances of the type you specified

- `map`: Returns a Flow that includes values from the Flow transformed with the operation you specified

- `mapNotNull`: Like map (transforms the Flow using the operation specified) but only includes values that are not null

- `withIndex`: Returns a Flow that converts each value to an **IndexedValue** containing the index of the value and the value itself

- `onEach`: Returns a Flow that performs the specified action on each value before they are emitted

- `runningReduce`: Returns a Flow containing the accumulated values resulting from running the operation specified sequentially, starting with the first element

- `runningFold`: Returns a Flow containing accumulated values resulting from running the operation specified sequentially, starting with the initial value set

- `scan`: Like the `runningFold` operator

There is also a `transform` operator that you can use to apply custom or complex operations. With the `transform` operator, you can emit values into the new Flow by calling the `emit` function with the value to send.

For example, this `MovieViewModel` has a `fetchTopMovieTitles` function that uses `transform` to return a Flow with the top movies:

```
class MovieViewModel : ViewModel() {

    ...

    fun fetchTopMovies(): Flow<Movie> {
        return fetchMoviesFlow()
            .transform {
                if (it.popularity > 0.5f) emit(it)
            }
    }

}
```

In this example, the `transform` operator was used in the Flow of movies to return a new Flow. The `transform` operator was used to emit only the list of movies whose popularity is higher than `0.5`, which means a popularity of more than 50%.

There are also size-limiting operators that you can use with Flow. The following are some of these operators:

- `drop(x)`: Returns a Flow that ignores the first *x* elements
- `dropWhile`: Returns a Flow that ignores the first elements that meet the condition specified
- `take(x)`: Returns a Flow containing the first *x* elements of the Flow
- `takeWhile`: Returns a Flow that includes the first elements that meet the condition specified

These size-limiting operators also function similarly to the Kotlin collection functions with the same name.

In this section, we learned about Intermediate flow operators. Intermediate operators transform a Flow into a new Flow. In the next section, we will learn how to buffer and combine Kotlin Flows.

Buffering and combining flows

In this section, we will learn about buffering and combining Kotlin Flows. You can buffer and combine Flows with Flow operators. Buffering allows Flow with long-running tasks to run independently and avoid race conditions. Combining allows you to join different sources of Flows before processing or displaying them on the screen.

Buffering Kotlin Flows

In this section, we will learn about buffering Kotlin Flows. Buffering allows you to run data emission in parallel to the collection.

Emitting and collecting data with Flow run sequentially. When a new value is emitted, it will be collected. Emission of a new value can only happen once the previous data has been collected. If the emission or the collection of data from the Flow takes a while to complete, the whole process will take a longer time.

With buffering, you can make a Flow's emission and collection of data run in parallel. There are three operators you can use to buffer Flows:

- `buffer`
- `conflate`
- `collectLatest`

`buffer()` allows the Flow to emit values while the data is still being collected. The emission and collection of data are run in separate coroutines, so it runs in parallel. The following is an example of how to use `buffer` with Flows:

```
class MainActivity : AppCompatActivity() {

    ...


    override fun onCreate(savedInstanceState: Bundle?) {

        ...

        lifecycleScope.launch {
            repeatOnLifecycle(Lifecycle.State.STARTED) {
                viewModel.fetchMovies()
                    .buffer()
                    .collect { movie ->
                        processMovie(movie)
                    }
            }
        }
    }
}
```

Here, the `buffer` operator was added before calling `collect`. If the `processMovie(movie)` function in the collection takes longer, the Flow will emit and buffer the values before they are collected and processed.

`conflate()` is similar to the `buffer()` operator, except with `conflate`, the collector will only process the latest value emitted after the previous value has been processed. It will ignore the other values previously emitted. Here is an example of using `conflate` in a Flow:

```
class MainActivity : AppCompatActivity() {

    ...


    override fun onCreate(savedInstanceState: Bundle?) {

        ...
        lifecycleScope.launch {
            repeatOnLifecycle(Lifecycle.State.STARTED) {
                viewModel.getTopMovie()
```

```
                            .conflate()
                            .collect { movie ->
                                processMovie(movie)
                            }
                        }
                    }
                }
            }
```

In this example, adding the `conflate` operator will allow us to only process the latest value from the Flow and call `processMovie` with that value.

`collectLatest(action)` is a terminal operator that will collect the Flow the same way as `collect`, but whenever a new value is emitted, it will restart the action and use this new value. Here is an example of using `collectLatest` in a Flow:

```
class MainActivity : AppCompatActivity() {

    ...

    override fun onCreate(savedInstanceState: Bundle?) {
        ...
        lifecycleScope.launch {
            repeatOnLifecycle(Lifecycle.State.STARTED) {
                viewModel.getTopMovie()
                    .collectLatest { movie ->
                        displayMovie(movie)
                    }
            }
        }
    }
}
```

Here, `collectLatest` was used instead of the `collect` terminal operator to collect the flow from `viewModel.getTopMovie()`. Whenever a new value is emitted by this Flow, it will restart and call `displayMovie` with the new value.

In this section, you learned how to buffer Kotlin Flows with `buffer`, `conflate`, and `collectLatest`. In the next section, you will learn about combining multiple Flows into a single Flow.

Combining Flows

In this section, we will learn how we can combine Flows. The Kotlin Flow API has available operators that you can use to combine multiple flows.

If you have multiple flows and you want to combine them into one, you can use the following Flow operators:

- `zip`
- `merge`
- `combine`

`merge` is a top-level function that combines the elements from multiple Flows of the same type into one. You can pass a `vararg` number of Flows to combine. This is useful when you have two or more sources of data that you want to merge first before collecting.

In the following example, there are two Flows from `viewModel.fetchMoviesFromDb` and `viewModel.fetchMoviesFromNetwork` combined using `merge`:

```
class MainActivity : AppCompatActivity() {

    ...

    override fun onCreate(savedInstanceState: Bundle?) {
        ...
        lifecycleScope.launch {
            repeatOnLifecycle(Lifecycle.State.STARTED) {
                merge(viewModel.fetchMoviesFromDb(),
                    viewModel.fetchMoviesFromNetwork())
                        .collect { movie ->
                            processMovie(movie)
                        }
            }
        }
    }
}
```

In this example, `merge` was used to combine the Flows from `viewModel.fetchMoviesFromDb` and `viewModel.fetchMoviesFromNetwork` before they are collected.

The `zip` operator pairs data from the first Flow to the second Flow into a new value using the function you specified. If one Flow has fewer values than the other, `zip` will end when the values of this Flow have all been processed.

The following shows how you can use the `zip` operator to combine two Flows, `userFlow` and `taskFlow`:

```
class MainActivity : AppCompatActivity() {
    ...

    override fun onCreate(savedInstanceState: Bundle?) {
        ...
        lifecycleScope.launch {
            repeatOnLifecycle(Lifecycle.State.STARTED) {
                val userFlow = viewModel.getUsers()
                val taskFlow = viewModel.getTasks()
                userFlow.zip(taskFlow) { user, task ->
                    AssignedTask(user, task)
                }.collect { assignedTask ->
                    displayAssignedTask(assignedTask)
                }
            }
        }
    }
}
```

In this example, you used `zip` to pair each value of `userFlow` to `taskFlow` and return a Flow of `AssignedTask` using the `user` and `task` values. This new Flow will be collected and then displayed with the `displayAssignedTask` function.

`combine` pairs data from the first flow to the second flow like `zip` but uses the most recent value emitted by each flow. It will continue to run as long as a Flow emits a value. There is also a top-level `combine` function that you can use for multiple flows.

The following example shows how you can use the `combine` operator to join two Flows in your application:

```
class MainActivity : AppCompatActivity() {
    ...

    override fun onCreate(savedInstanceState: Bundle?) {
```

```
        . . .
        lifecycleScope.launch {
            repeatOnLifecycle(Lifecycle.State.STARTED) {
                val yourMesssage =
                    viewModel.getLastMessageSent()
                val friendMessage =
                    viewModel.getLastMessageReceived()
                userFlow.combine(taskFlow) { yourMesssage,
                    friendMessage ->
                        Conversation(yourMessage,
                            friendMessage)
                }.collect { conversation ->
                    displayConversation(conversation)
                }
            }
        }
    }
}
```

Here, you have two Flows, `yourMessage` and `friendMessage`. The `combine` function pairs the most recent value of `yourMessage` and `friendMessage` to create a `Conversation` object. Whenever a new value is emitted by either Flow, `combine` will pair the latest values and add that to the resulting Flow for collection.

In this section, we have explored how to combine Flows. In the next section, we will focus on `StateFlow` and `SharedFlow` and how we can use them in your Android applications.

Exploring StateFlow and SharedFlow

In this section, we will dive into `StateFlow` and `SharedFlow`. `SharedFlow` and `StateFlow` are Flows that are hot streams, unlike a normal Kotlin Flow, which are cold streams by default.

A Flow is a cold stream of data. Flows only emit values when the values are collected. With `SharedFlow` and `StateFlow` hot streams, you can run and emit values the moment they are called and even when they have no listeners. `SharedFlow` and `StateFlow` are Flows, so you can also use operators on them.

A `SharedFlow` allows you to emit values to multiple listeners. `SharedFlow` can be used for one-time events. The tasks that will be done by the `SharedFlow` will only be run once and will be shared by the listeners.

You can use `MutableSharedFlow` and then use the `emit` function to send values to all the collectors.

In the following example, `SharedFlow` is used in `MovieViewModel` for the list of movies fetched:

```
class MovieViewModel : ViewModel() {
    private val _message = MutableSharedFlow<String>()
    val movies: SharedFlow<String> =
      _message.asSharedFlow()
    ...


    fun onError(): Flow<List<Movie>> {
        ...
        _message.emit("An error was encountered")
    }
}
```

In this example, we used `SharedFlow` for the message. We used the `emit` function to send the error message to the Flow's listeners.

`StateFlow` is `SharedFlow`, but it only emits the latest value to its listeners. `StateFlow` is initialized with a value (an initial state) and keeps this state. You can change the value of `StateFlow` using the mutable version of `StateFlow`, `MutableStateFlow`. Updating the value sends the new value to the Flow.

In Android, `StateFlow` can be an alternative to `LiveData`. You can use `StateFlow` for `ViewModel`, and your activity or fragment can then collect the value. For example, in the following `ViewModel`, `StateFlow` is used for the list of movies:

```
class MovieViewModel : ViewModel() {
    private val _movies =
      MutableStateFlow(emptyList<Movie>())
    val movies: StateFlow<List<Movie>> = _movies
    ...


    fun fetchMovies(): Flow<List<Movie>> {
        ...
        _movies.value = movieRepository.fetchMovies()
    }
}
```

In the preceding code, the list of movies fetched from the repository will be set to `MutableStateFlow` of _movies, which will also change `StateFlow` of movies. You can then collect `StateFlow` of movies in an activity or fragment, as shown in the following:

```
class MainActivity : AppCompatActivity() {
    ...

    override fun onCreate(savedInstanceState: Bundle?) {
        ...
        lifecycleScope.launch {
            repeatOnLifecycle(Lifecycle.State.STARTED) {
                viewModel.movies.collect { movies ->
                    displayMovies(movies)
                }
            }
        }
    }
}
```

Here, `StateFlow` of `viewModel.movies` will be collected, and then the list of movies will be displayed on the screen with the `displayMovies` function.

In this section, we have learned about `StateFlow` and `SharedFlow` and how we can use them in our Android projects.

Let's try what we have learned so far by adding Kotlin Flow to an Android project.

Exercise 5.01 – Using Kotlin Flow in an Android app

For this exercise, you will be continuing the movie app you worked on in *Exercise 4.01 – Adding tests to coroutines in an Android app*. This application displays the movies that are currently playing in cinemas. You will be adding Kotlin Flow to the project by following these steps:

1. Open the movie app you worked on in *Exercise 4.01 – Adding tests to coroutines in an Android app* in Android Studio.

2. Go to the `MovieRepository` class and add a new `fetchMoviesFlow()` function that uses a `flow` builder to return a Flow and emits the list of movies from `MovieService`, as shown in the following snippet:

```
fun fetchMoviesFlow(): Flow<List<Movie>> {
    return flow {
```

```
            emit(movieService.getMovies(apiKey).results)
        }.flowOn(Dispatchers.IO)
    }
```

This is the same as the `fetchMovies()` function, but this function uses Kotlin Flow and will return `Flow<List<Movie>>` to the function or class that will collect it. The Flow will emit the list of movies from `movieService.getMovies`, and it will flow on the `Dispatchers.IO` dispatcher.

3. Open the `MovieViewModel` class, and replace the initialization of the `movies` `LiveData` that gets the value from `movieRepository` with the following lines:

```
private val _movies =
    MutableStateFlow(emptyList<Movie>())
val movies: StateFlow<List<Movie>> = _movies
```

This will allow you to use the value of the `_movies` `MutableStateFlow` as the value of the `movies` `StateFlow`, which you will change later when you have fetched the list of movies from the Flow in `movieRepository`.

4. Do the same for the `error` `LiveData`, and replace its initialization with the value from `movieRepository` with the following lines:

```
private val _error = MutableStateFlow("")
val error: StateFlow<String> = _error
```

This will use the value of the `_error` `MutableStateFlow` for the `error` `StateFlow`. You will be able to change the value of this `StateFlow` later for handling the cases when the Flow encountered an exception.

5. Replace the `loading` and `_loading` variables with the following lines:

```
private val _loading = MutableStateFlow(true)
val loading: StateFlow<String> = _loading
```

This will use the value of the `_loading` `MutableStateFlow` for the `loading` `StateFlow`. You will update this later to indicate that the loading of movies is ongoing.

6. Remove the `fetchMovies()` function and its content. You will be replacing this in the next step.

7. Add a new `fetchMovies()` function that will collect the Flow from the `movieRepository.fetchMoviesFlow`, as shown in the following code block:

```
fun fetchMovies() {
    _loading.value = true
    viewModelScope.launch (dispatcher) {
```

```
MovieRepository.fetchMoviesFlow()
    .collect {
        _movies.value = it
        _loading.value = false
    }
}
}
```

This will collect the list of movies from movieRepository.fetchMoviesFlow and set it to the _movies MutableStateFlow and the movies StateFlow. This list of movies will then be displayed in MainActivity.

8. Open the app/build.gradle file. Add the following lines in the dependencies:

```
implementation 'androidx.lifecycle:lifecycle-runtime-
ktx:2.4.1'
```

This will allow us to use lifecycleScope for collecting the flows in MainActivity later.

9. Open MainActivity and remove the lines of code that observe for the movies, error, and loading LiveData. Replace them with the following:

```
lifecycleScope.launch {
    repeatOnLifecycle(Lifecycle.State.STARTED) {
        launch {
            movieViewModel.movies.collect { movies ->
                movieAdapter.addMovies(movies)
            }
        }
        launch {
            movieViewModel.error.collect { error ->
                if (error.isNotEmpty())
                    Snackbar.make(recyclerView, error,
                    Snackbar.LENGTH_LONG).show()
            }
        }
        launch {
            movieViewModel.loading.collect { loading ->
                progressBar.isVisible = loading
            }
```

```
                }
            }
        }
```

This will collect `movies` and add them to the list, collect the `error` and display a `SnackBar` message if `error` is not empty, and collect `loading` and update `progressBar` based on its value.

10. Run the application. The app should still display a list of movies (with a poster and a title), as shown in the following screenshot:

Figure 5.1 – The movie app with the list of movies

In this exercise, we have added Kotlin Flow in an Android app by creating a `MovieRepository` function that returns the list of movies as a Flow. This Flow was then collected by `MovieViewModel`.

Summary

This chapter focused on using Kotlin Flows for asynchronous programming in Android. Flows are built on top of Kotlin coroutines. A flow can emit multiple values sequentially, instead of just a single value.

We started with learning about how to use Kotlin Flows in your Android app. Jetpack libraries such as Room and some third-party libraries support Flow. To safely collect flows in the UI layer and prevent memory leaks and avoid wasting resources, you can use `Lifecycle.repeatOnLifecycle` and `Flow.flowWithLifecycle`.

We then moved on to creating Flows with Flow builders. The `flowOf` function creates a Flow that emits the value or `vararg` values you provided. You can convert collections and functional types to Flow with the `asFlow()` extension function. The `flow` builder function creates a new Flow from a suspending lambda block, inside which you can send values with `emit()`.

Then, we explored Flow operators and learned how you can use them with Kotlin Flows. With terminal operators, you can start the collection of the Flow. Intermediate operators allow you to transform a Flow into another Flow.

We then learned about buffering and combining Flows. With the `buffer`, `conflate`, and `collectLatest` operators, you can buffer Flows. You can combine Flows with the `merge`, `zip`, and `combine` Flow operators.

We then explored `SharedFlow` and `StateFlow`. These can be used in your Android projects. With `SharedFlow`, you can emit values to multiple listeners. `StateFlow` is `SharedFlow` that only emits the latest value to its listeners.

Finally, we worked on an exercise to add Kotlin Flows to an Android application. We used a Flow in `MovieRepository`, which was then collected in `MovieViewModel`.

In the next chapter, we will focus on how to handle Kotlin Flows cancelations and exceptions in your application.

6

Handling Flow Cancelations and Exceptions

In the previous chapter, we focused on Kotlin Flows and learned how we can use them in our Android projects. We learned about creating Kotlin Flows with Flow builders. We then explored Flow operators and how to use them with Kotlin Flows. We then learned about buffering and combining Flows. Finally, we explored `SharedFlow` and `StateFlow`.

Flows can be canceled, and they can fail or encounter exceptions. Developers must be able to handle these properly to prevent application crashes and to inform their users with a dialog or a toast message. We will discuss how to do these tasks in this chapter.

In this chapter, we will start by understanding Flow cancelation. We will learn how to cancel Flows and handle cancelations for our Flows. Then, we will learn how to manage failures and exceptions that can happen in our Flows. We will also learn about retrying and handling Flow completion.

In this chapter, we are going to cover the following main topics:

- Canceling Kotlin Flows
- Retrying tasks with Flow
- Catching exceptions in Flows
- Handling flow completion

By the end of this chapter, you will understand how to cancel flows, and will have learned how to manage cancelations and how to handle exceptions in flows.

Technical requirements

You will need to download and install the latest version of Android Studio. You can find the latest version at `https://developer.android.com/studio`. For an optimal learning experience, a computer with the following specifications is recommended:

- Intel Core i5 or equivalent or higher
- 4 GB RAM minimum
- 4 GB available space

The code examples for this chapter can be found on GitHub at `https://github.com/PacktPublishing/Simplifying-Android-Development-with-Coroutines-and-Flows/tree/main/Chapter06`

Canceling Kotlin Flows

In this section, we will start by looking at Kotlin Flow cancelations. Like coroutines, Flows can also be canceled manually or automatically.

In *Chapter 3, Handling Coroutine Cancelations and Exceptions*, we learned about canceling coroutines and that coroutine cancellation must be cooperative. As Kotlin Flows are built on top of coroutines, Flow follows the cooperative cancellation of coroutines.

Flows created using the `flow{}` builder are cancellable by default. Each `emit` call to send new values to the Flow also calls `ensureActive` internally. This checks whether the coroutine is still active, and if not, it will throw `CancellationException`.

For example, we can use the `flow{}` builder to create a cancellable Flow, as shown in the following:

```
class MovieViewModel : ViewModel() {

    ...

    fun fetchMovies(): Flow<Movie> = flow {
        movieRepository.fetchMovies().forEach {
            emit(it)
        }
    }
}
```

In the `fetchMovies` function here, we used the `flow` builder to create the Flow of movies returned by `movieRepository.fetchMovies`. This `Flow<Movie>` will be a cancellable Flow by default.

All other Flows, such as ones created using the `asFlow` and `flowOf` Flow builders, are not cancellable by default. We must handle the cancellation ourselves. There is a `cancellable()` operator we can use on a Flow to make it cancelable. This will add an `ensureActive` call on each emission of a new value.

The following example shows how we can make a Flow cancelable using the `cancellable` Flow operator:

```
class MovieViewModel : ViewModel() {

    ...


    fun fetchMovies(): Flow<Movie> {
        return movieRepository.fetchMovies().cancellable()
    }
}
```

In this example, we used the cancelable operator on the Flow returned by `movieRepository. fetchMovies()` to make the resulting Flow cancelable.

In this section, we learned how to cancel Kotlin Flows and how to make sure your Flows can be cancellable. In the next section, we will focus on how to retry your tasks with Kotlin Flows.

Retrying tasks with Flow

In this section, we will explore Kotlin Flow retrying. There are cases when retrying an operation is needed for your application.

When performing long-running tasks, such as a network call, sometimes it is necessary to try the call again. This includes cases such as logging in/out, posting data, or even fetching data. The user may be in an area with a low internet connection, or there may be other factors why the call is failing. With Kotlin Flows, we have the `retry` and `retryWhen` operators that we can use to retry Flows automatically.

The `retry` operator allows you to set a **Long** `retries` as the maximum number of times the Flow will retry. You can also set a `predicate` condition, a code block that will retry the Flow when it returns `true`. The predicate has a **Throwable** parameter representing the exception that occurred; you can use that to check whether you want to do the retry or not.

The following example shows how we can use the `retry` Flow operator to retry our tasks in our Flow:

```
class MovieViewModel : ViewModel() {
    ...
```

```
fun favoriteMovie(id: Int) =
    movieRepository.favoriteMovie(id)
        .retry(3) { cause -> cause is IOException }
}
```

Here, the Flow from `movieRepository.favoriteMovie(id)` will be retried up to three times when the exception encountered is `IOException`.

If you do not pass a value for the retries, the default of `Long.MAX_VALUE` will be used. `predicate`, when not provided, has a default value of `true`, meaning the Flow will always be retried if `retries` has not yet been reached.

The `retryWhen` operator is similar to the `retry` operator. We need to specify `predicate`, which is the condition and only when `true` will it perform the retry. `predicate` has a **Throwable** parameter representing the exception encountered and a **Long** parameter for the number of attempts (which starts at zero). We can use both to create the condition which, if evaluated to `true`, will retry the Flow. The following code shows an example of using `retryWhen` to retry your tasks in your Flow:

```
class MovieViewModel : ViewModel() {

    ...

    fun favoriteMovie(id: Int) =
        movieRepository.favoriteMovie(id)
            .retryWhen { cause, attempt -> attempt <3 &&
                cause is IOException }
}
```

In this example, we used `retryWhen` and specified that the retry will be done when the value of `attempt` is less than three and only if the exception is `IOException`.

With the `retryWhen` operator, we can also emit a value to the Flow (with the `emit` function), which we can use to represent the retry attempt or a value. We can then display this value on the screen or process it. The following example shows how we can use `emit` with the `retryWhen` operator:

```
class MovieViewModel : ViewModel() {

    ...

    fun getTopMovieTitle(): Flow<String> {
        return movieRepository.getTopMovieTitle(id)
            .retryWhen { cause, attempt ->
                emit("Fetching title again...")
                attempt <3 && cause is IOException
```

```
        }
        ...
}
```

Here, the Flow's task will be retried when the number of attempts is less than three and only if the exception is **IOException**. It will then emit a `Fetching title again` string that can be processed by the activity or fragment that listens to the Flow returned by `MovieViewModel.getTopMovieTitle()`.

In this section, you learned about retrying tasks such as network requests with Kotlin Flow. We will explore Kotlin Flow exceptions and how to update our code to catch these exceptions in the next section.

Catching exceptions in Flows

The Flows in your code can encounter `CancellationException` when they are canceled or other exceptions when emitting or collecting values. In this section, we will learn how to handle these Kotlin Flow exceptions.

Exceptions can happen in Flows during the collection of values or when using any operators on a Flow. We can handle exceptions in Flows by enclosing the collection of the Flow in our code with a `try-catch` block. For example, in the following code, the `try-catch` block is used to add exception handling:

```kotlin
class MainActivity : AppCompatActivity() {

  ...

  override fun onCreate(savedInstanceState: Bundle?) {
      ...
      lifecycleScope.launch {
          repeatOnLifecycle(Lifecycle.State.STARTED) {
              try {
                  viewModel.fetchMovies().collect { movie ->
                      processMovie(movie)
                  }
              } catch (exception: Exception) {
                  Log.e("Error", exception.message)
              }
          }
      }
  }
}
```

Here, the collection code for the Flow returned by `viewModel.fetchMovies` was wrapped in a `try-catch` block. If an exception was encountered in the Flow, the exception message will be logged with the `Error` tag and `exception.message` as the message.

We can also use the `catch` Flow operator to handle exceptions in our Flow. With the `catch` operator, we can catch the exceptions from the upstream Flow, or the function and operators before the `catch` operator was called.

In the following example, the `catch` operator was used to catch exceptions from the Flow returned by `viewModel.fetchMovies`:

```
class MainActivity : AppCompatActivity() {
  ...

  override fun onCreate(savedInstanceState: Bundle?) {
      ...
      lifecycleScope.launch {
          repeatOnLifecycle(Lifecycle.State.STARTED) {
              viewModel.fetchMovies()
                  .catch { exception ->
                      handleException(exception) }
                  .collect { movie -> processMovie(movie) }
          }
      }
  }
}
```

Here, the `catch` operator was used in the Flow to catch the exceptions. The exception, which is an instance of **Throwable**, was then passed to the `handleException` function that is going to handle the exception.

We can also use the `catch` operator to emit a new value to represent the error or for use as a fallback value instead, such an empty list. In the following example, a default string value of `No Movie Fetched` will be used when an exception occurs in the Flow that returns the title of the top movie:

```
class MainActivity : AppCompatActivity() {
  ...

  override fun onCreate(savedInstanceState: Bundle?) {
      ...
      lifecycleScope.launch {
```

```
        repeatOnLifecycle(Lifecycle.State.STARTED) {
            viewModel.getTopMovieTitle()
                .catch { emit("No Movie Fetched") }
                .collect { title -> displayTitle(title) }
        }
    }
}
}
```

In this example, we used the catch operator to emit the No Movie Fetched string when an exception occurs in getting the top movie title from ViewModel. This will be the value that will be used in the displayTitle() call.

As the catch operator only handles exceptions in the upstream Flow, an exception that happens during the collect{} call won't be caught. While you can use the try-catch block to handle these exceptions, you can also move the collection code to an onEach operator, add the catch operator after it, and use collect() to start the collection.

The following example shows how your code can look when using an onEach operator for the collection of values and the catch operator for handling exceptions:

```
class MainActivity : AppCompatActivity() {

    ...

    override fun onCreate(savedInstanceState: Bundle?) {

        ...

        lifecycleScope.launch {
            repeatOnLifecycle(Lifecycle.State.STARTED) {
                viewModel.fetchMovies()
                    .onEach { movie -> processMovie(movie) }
                    .catch { exception ->
                        handleError(exception) }
                    .collect()
            }
        }
    }
}
```

Here, the collect() function without parameters was used, and the onEach operator will process each movie from the Flow.

In this section, we learned how to catch exceptions in Flows. In the following section, we will focus on Kotlin Flow completion.

Handling Flow completion

In this section, we will explore how to handle Flow completion. We can add code to perform additional tasks after our Flows have completed.

When the Flow encounters an exception, it will be canceled and complete the Flow. A Flow is also completed when the last element of the Flow has been emitted.

To add a listener in your Flow when it has completed, you can use the `onCompletion` operator and add the code block that will be run when the Flow completes. A common usage of `onCompletion` is hiding the **ProgressBar** in your UI when the Flow has completed, as shown in the following code:

```kotlin
class MainActivity : AppCompatActivity() {
    ...

    override fun onCreate(savedInstanceState: Bundle?) {
        ...
        lifecycleScope.launch {
            repeatOnLifecycle(Lifecycle.State.STARTED) {
                viewModel.fetchMovies()
                    .onStart { progressBar.isVisible = true }
                    .onEach { movie -> processMovie(movie) }
                    .onCompletion { progressBar.isVisible =
                        false }
                    .catch { exception ->
                        handleError(exception) }
                    .collect()
            }
        }
    }
}
```

In this example, we have added the `onCompletion` operator to hide `progressBar` when the Flow has completed. We have also used `onStart` to display `progressBar`.

The `onStart` operator is the opposite of `onCompletion`. It will be called before the Flow starts emitting values. In the previous example, `onStart` was used so that before the Flow starts, `progressBar` will be displayed on the screen.

Within the code block you add in onStart and onCompletion (if the Flow completed successfully and without exception), you can also emit values, such as an initial and final value. In the following example, an onStart operator is used to emit an initial value to be displayed on the screen:

```
class MainActivity : AppCompatActivity() {
  ...


  override fun onCreate(savedInstanceState: Bundle?) {
    ...
    lifecycleScope.launch {
      repeatOnLifecycle(Lifecycle.State.STARTED) {
        viewModel.getTopMovieTitle()
          .onStart { emit("Loading...") }
          .catch { emit("No Movie Fetched") }
          .collect { title -> displayTitle(title) }
      }
    }
  }
}
```

Here, onStart is used to listen to when the Flow starts. When the Flow starts, it will emit a Loading... string as the initial value of the Flow. This will then be the first item that will be displayed on the screen.

The onCompletion code block also has a nullable **Throwable** that corresponds to the exception thrown by the Flow. It will be null if the Flow has completed successfully. However, unlike catch, the exception itself will not be handled, so you still need to use catch or try-catch to handle this exception.

The following example shows how we can use this nullable **Throwable** in our Flow's onCompletion call:

```
class MainActivity : AppCompatActivity() {
  ...


  override fun onCreate(savedInstanceState: Bundle?) {
    ...
    lifecycleScope.launch {
      repeatOnLifecycle(Lifecycle.State.STARTED) {
        viewModel.getTopMovieTitle()
          .onCompletion { cause ->
```

```
                    progressBar.isVisible = false
                    if (cause != null) displayError(cause)
                }
                .catch { emit("No Movie Fetched") }
                .collect { title -> displayTitle(title) }
            }
        }
    }
}
```

In this example, we checked the cause in the onCompletion block, and if it's not null (which means an exception was encountered), displayError will be called and the cause passed to it.

In this section, we learned about onStart and onCompletion to handle when Flows start and when they are completed.

Let's try what you have learned by adding code to handle exceptions that can occur in Flows in an Android project.

Exercise 6.01 – Handling Flow exception in an Android app

In this exercise, you will be continuing with the movie app you worked on in *Exercise 5.01 – Using Kotlin Flow in an Android app*. This application displays the movies that are playing now in movie theaters. You will be updating the project to handle Flow cancelations and exceptions by following these steps:

1. In Android Studio, open the movie app you worked on in *Exercise 5.01 – Using Kotlin Flow in an Android app*.

2. Go to the MovieViewModel class. In the fetchMovies function, remove the line that sets the value of _loading to true. Your function will look like the following:

```
fun fetchMovies() {
    viewModelScope.launch (dispatcher) {
        MovieRepository.fetchMoviesFlow()
            .collect {
                _movies.value = it
                _loading.value = false
            }
    }
}
```

You removed the code that sets `loading` to `true` (and displays `ProgressBar` on the screen). It will be replaced in the next step with an `onStart` Flow operator.

3. Add an `onStart` operator before the `collect` call, which will set the value of `_loading` to `true` when the Flow starts, as shown in the following:

```
fun fetchMovies() {
    viewModelScope.launch (dispatcher) {
        MovieRepository.fetchMoviesFlow()
            .onStart { _loading.value = true }
            .collect {
                _movies.value = it
                _loading.value = false
            }
    }
}
```

The `onStart` operator will set the value of `_loading` to `true` and display `ProgressBar` on the screen when the Flow starts.

4. Next, remove the line that sets the value of `_loading` to `false` in the code block inside the `collect` call. Your function will look like the following:

```
fun fetchMovies() {
    viewModelScope.launch (dispatcher) {
        MovieRepository.fetchMoviesFlow()
            .onStart { _loading.value = true }
            .collect {
                _movies.value = it
            }
    }
}
```

You removed the code that sets the value of `_loading` to `false` and hides `ProgressBar` on the screen when the Flow is collected.

5. Add an `onCompletion` operator before the `collect` call, which will set the value of `_loading` to `false` when the Flow has completed, as shown in the following:

```
fun fetchMovies() {
    viewModelScope.launch (dispatcher) {
        MovieRepository.fetchMoviesFlow()
            .onStart { _loading.value = true }
```

```
        .onCompletion { _loading.value = false }
        .collect {
            _movies.value = it
        }
    }
}
```

The `onCompletion` Flow operator will set the value of `_loading` to `false`. This will then hide, upon completion of the Flow, `ProgressBar`, which is displayed on the screen while the movies are being fetched.

6. Add a `catch` operator before the `collect` function to handle the case when the Flow has encountered an exception:

```
fun fetchMovies() {
    viewModelScope.launch (dispatcher) {
        MovieRepository.fetchMoviesFlow()
            .onStart { _loading.value = true }
            .onCompletion { _loading.value = false }
            .catch {
                _error.value = "An exception occurred:
                    ${it.message}"
            }
            .collect {
                _movies.value = it
            }
    }
}
```

This will set a string containing An exception occurred: and the exception message as the value of the `_error` LiveData. This `_error` LiveData will display an error message in `MainActivity`.

7. On your device or emulator, turn off the Wi-Fi and mobile data. Then, run the app. This will cause an error in fetching the movies, as there is no internet connection. The app will display a `SnackBar` message, as shown in the following screenshot:

Figure 6.1 – The error message displayed in the movie app

8. Close the application and turn on the Wi-Fi and/or mobile data on your device or emulator. Run the application again. The app should show `ProgressBar`, display a list of movies (with the movie title and poster) on the screen, and hide `ProgressBar`, as shown in the following screenshot:

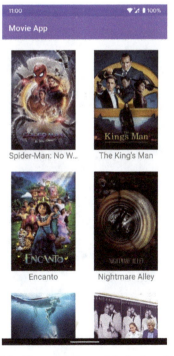

Figure 6.2 – The movie app with the list of movies

In this exercise, you have updated the application so that it can handle exceptions in the Flow instead of crashing.

Summary

This chapter focused on Kotlin Flow cancelations. You learned that Flows follow the cooperative cancellation of coroutines. The `flow{ }` builder and `StateFlow` and `SharedFlow` implementations are cancellable by default. You can use the `cancellable` operator to make other Flows cancellable.

We then learned about retrying tasks with Kotlin Flow. You can use the `retry` and `retryWhen` functions to retry the Flow based on the number of attempts and the exception encountered by the Flow.

Then, we learned about handling exceptions that can happen during the emission or collection of data in a Flow. You can use the `try-catch` block or the `catch` Flow operator to handle Flow exceptions.

We learned how to handle Flow completion. With the `onStart` and `onCompletion` operators, you can listen and run code when Flows start and when they have finished. You can also emit values with the `onStart` and `onCompletion` code blocks, such as when you want to set an initial and final value for the Flow.

Finally, we worked on an exercise to update our Android project and handle the exceptions that can be encountered in a Flow. We used the `catch` Flow operator to handle exceptions in the project.

In the next chapter, we will dive into creating and running tests for the Kotlin Flows in our Android projects.

7

Testing Kotlin Flows

In the previous chapter, we focused on understanding Kotlin Flow cancellation, learning how to make Flows cancellable, and handling the cancellation. We also learned about retrying tasks with Flows and handling completion and exceptions in your Flows.

Adding tests for the Kotlin Flows in your code is an important part of app development. Tests will ensure that the Flows we add to our projects are free of bugs or errors and that they will work as we intended. They can make developing apps easier and help you refactor and maintain your code confidently.

In this chapter, we will learn how to test Kotlin Flows in Android. First, we will understand how to set up your Android project for testing Flows. We will then proceed with creating and running tests for Kotlin Flows.

This chapter covers the following main topics:

- Setting up an Android project for testing Flows
- Testing Kotlin Flows
- Testing Flows with Turbine

By the end of this chapter, you will have learned about Kotlin Flow testing. You will be able to write and run unit and integration tests for the Flows in your Android applications.

Technical requirements

You will need to download and install the latest version of Android Studio. You can find the latest version at https://developer.android.com/studio. For an optimal learning experience, a computer with the following specifications is recommended:

- Intel Core i5 or equivalent or higher
- A minimum of 4 GB of RAM
- 4 GB of available space

The code examples for this chapter can be found on GitHub at `https://github.com/PacktPublishing/Simplifying-Android-Development-with-Coroutines-and-Flows/tree/main/Chapter07`.

Setting up an Android project for testing Flows

In this section, we will start by looking at how to set up our Android project for testing Kotlin Flows. Once we have done that, it will be easy for us to add unit and integration tests for the Flows in our project.

To create a unit test in Android, your project must have the JUnit 4 testing library, a unit testing framework for Java. New projects created in Android Studio should already have this added in the `app/build` dependencies. If your project does not have JUnit yet, you can add it by adding the following in your `app/build.gradle` dependencies:

```
dependencies {
    …
    testImplementation 'junit:junit:4.13.2'
}
```

Adding this to your dependencies enables you to use the JUnit 4 testing framework to unit-test your code.

It is also a good idea to use mock objects for your tests. Mockito is a popular Java mocking library that you can use on Android. You can also use Mockito-Kotlin to use Mockito with idiomatic Kotlin code. To add Mockito and Mockito-Kotlin to your Android tests, you can add the following in your `app/build.gradle` dependencies:

```
dependencies {
    …
    testImplementation 'org.mockito:mockito-core:4.0.0'
    testImplementation 'org.mockito.kotlin:mockito-
        kotlin:4.0.0'
}
```

This will allow you to use Mockito to create mock objects for your Android tests using Kotlin-like code. Mockito-Kotlin has a dependency to **mockito-core** so you can simply use the following to import both `mockito-core` and `mockito-kotlin`:

```
dependencies {
    …
```

```
        testImplementation 'org.mockito.kotlin:mockito-
          kotlin:4.0.0'
}
```

As Kotlin Flow is built on top of coroutines, you can use the `kotlinx-coroutines-test` library to help you add tests for both coroutines and Flows. This library contains utility classes to make the writing of tests easier. To add it to your project, you can add the following to your `app/build.gradle` dependencies:

```
dependencies {
    ...
    testImplementation 'org.jetbrains.kotlinx:kotlinx-
      coroutines-test:1.6.0'
}
```

Adding this allows you to use the `kotlinx-coroutines-test` library for testing coroutines and flows in your project.

In this section, we have learned about setting up our Android project to test Kotlin Flows. We will learn about testing Kotlin Flows in the next section.

Testing Kotlin Flows

In this section, we will focus on testing Kotlin Flows. We can create unit and integration tests for classes such as `ViewModel` that use Flow in their code.

To test code that collects a Flow, you can use a mock object that can return values which you can do assertion checks. For example, if your `ViewModel` listens to the Flow from a repository, you can create a custom `Repository` class that emits a Flow with a predefined set of values for easier testing.

For example, say you have a `MovieViewModel` class such as the following that has a `fetchMovies` function that collects a Flow:

```
class MovieViewModel(private val movieRepository:
  MovieRepository) {
    ...
    suspend fun fetchMovies() {
        movieRepository.fetchMovies().collect {
            _movies.value = it
        }
```

```
        }
    }
```

Here, the `fetchMovies` function collects a Flow from `movieRepository.fetchMovies()`. You can write a test for this `MovieViewModel` by creating `MovieRepository`, which returns a specific set of values, which you will check to see whether it's the same value that will be set to the movies `LiveData` in `MovieViewModel`. An example implementation of this looks like the following:

```
class MovieViewModelTest {
    ...

    @Test
    fun fetchMovies() {
        ...
        val list = listOf(movie1, movie2)
        val expected = MutableLiveData<List<Movie>>()
        expectedMovies.value = list

        val movieRepository: MovieRepository = mock {
            onBlocking { fetchMoviesFlow() } doReturn
              flowOf(movies)
        }
        val dispatcher = StandardTestDispatcher()
        val movieViewModel =
          MovieViewModel(movieRepository, dispatcher)

        runTest {
            movieViewModel.fetchMovies()
            dispatcher.scheduler.advanceUntilIdle()
            assertEquals(expectedMovies.value,
              movieViewModel.movies.value)
            ...
        }
    }
}
```

In this example, `fetchMoviesFlow` of `MovieRepository` returns a list of movies that has only one item. After calling `movieViewModel.fetchMovies()`, the test checks whether the value in the `MovieViewModel.movies LiveData` was set to this list.

You can also test a Flow by collecting it to another object. You can do that by converting the Flow to a list with `toList()` or to a set with `toSet()`, getting the first item with first, taking items with `take()`, and other terminal operators. Then, you can check the values returned with the expected values.

For example, say you have `MovieViewModel`, which has a function that returns a Flow, such as the following class:

```
class MovieViewModel(private val movieRepository:
  MovieRepository) {

    ...

    fun fetchFavoriteMovies(): Flow<List<Movie>> {

        ...

    }
}
```

Here, the `fetchFavoriteMovies` function returns a Flow of `List<Movie>`. You can write a test for this function by converting `Flow<List<Movie>>` into a list, as shown in the following example:

```
class MovieViewModelTest {

    ...

    @Test
    fun fetchFavoriteMovies() {

        ...
        val expectedList = listOf(movie1, movie2)

        val movieRepository: MovieRepository = mock {
            onBlocking { fetchFavoriteMovies() } doReturn
              flowOf(expectedList)
        }
        val movieViewModel =
          MovieViewModel(movieRepository)
```

```
        runTest {
            ...
            assertEquals(expectedList,
              movieViewModel.fetchFavoriteMovies().toList())
        }
    }
}
```

In this example, you converted the Flow of the list of movies from `movieViewModel.fetchFavoriteMovies()` to a list of movies and compared it with the expected list.

To test error-handling in Flow, you can mock your test objects to throw an exception. You can then check the exception thrown or the code that handles it. The following example shows how you can write tests for a Flow's failure case:

```
class MovieRepositoryTest {

    ...

    @Test
    fun fetchMoviesFlowWithError() {
        val movieService: MovieService = mock {
            onBlocking { getMovies(anyString()) } doThrow
              IOException(exception)
        }

        val movieRepository = MovieRepository(movieService)

        runTest {
            movieRepository.fetchMoviesFlow().catch {
                assertEquals(exception, it.message)
            }
        }
    }
}
```

In this test class, every time `MovieService.getMovies()` is called, it will throw `IOException`. We then call `movieRepository.fetchMoviesFlow()` and use the `catch` operator to handle the exception. Then, we compare the exception message with the expected string.

We can also test Flow retries by mocking our class to return a specific exception that would trigger a retry. For retries that still fail afterward, you can check the exception or the exception handling. To test retries that succeed, you can mock your class to either throw an exception or return a Flow that you can compare with the expected values.

The following example shows how you can test a Flow that has a retry for `IOException` and any number of attempts:

```
class MovieViewModelTest {

    ...

    @Test

    fun fetchMoviesWithError() {
        ...
        val movies = listOf(Movie(title = "Movie"))
        val exception = "Exception"
        val hasRetried = false

        val movieRepository: MovieRepository = mock {
            onBlocking { fetchMoviesFlow() } doAnswer {
                flow {
                    if (hasRetried) emit(movies) else throw
                        IOException (exception)
                }
            }
        }
        ...
    }
}
```

Here, we used a `hasRetried` variable to determine whether to return a Flow of movies or to throw an exception that can trigger a retry. It is `false` by default to allow a retry. Later in the code, we can change this value to `true` to return a Flow of movies, which we can then compare to the expected values.

In this section, we learned how to create and run tests for Kotlin Flows in our Android project. We will learn about testing hot flows with Turbine in the next section.

Testing Flows with Turbine

In this section, we will learn how to test Flows using Turbine, which is a third-party library that we can use to test flows in our project.

Hot flows such as `SharedFlow` and `StateFlow`, as you learned in the previous chapter, emit values even if there are no listeners. They also keep emitting values and do not complete. Testing them is a bit more complicated. You won't be able to convert these flows to a list and then compare it to the expected values.

To test hot flows and make testing other Flows easier, you can use a library from Cash App called Turbine (`https://github.com/cashapp/turbine`). Turbine is a small testing library for Kotlin Flow that you can use in Android.

You can use the Turbine testing library in your Android project by adding the following to your `app/build.gradle` dependencies:

```
dependencies {
    ...
    testImplementation 'app.cash.turbine:turbine:0.8.0'
}
```

Adding this will allow you to use the Turbine testing library in your project to test the Flow in your code.

Turbine has a `test` extension function on Flow. It has a suspending validation block, where you can consume items from the Flow one by one and compare them with the expected values. It will then cancel the Flow at the end of the validation block.

An example of using Turbine and the `test` extension function to test Flows is shown in the following code block:

```
class MovieViewModelTest {
    ...

    @Test
    fun fetchMovies() {
        ...
        val expectedList = listOf(movie1, movie2)


        val movieRepository: MovieRepository = mock {
            onBlocking { fetchMovies() } doReturn
```

```
                    flowOf(expectedList)
            }
        val movieViewModel =
          MovieViewModel(movieRepository)

        runTest {
            movieViewModel.fetchMovies().test {
                assertEquals(movie1, awaitItem())
                assertEquals(movie2, awaitItem())
                awaitComplete()
            }
        }
    }
}
```

Here, the test used an `awaitItem()` function to get the next item emitted by the Flow and compared it with the expected items. Then, it used an `awaitComplete()` function to assert that the Flow had completed.

To test for exceptions thrown by the Flow, you can use the `awaitError()` function that returns `Throwable`. You can then compare this `Throwable` to the one you expected to be thrown. The following example shows how you can use this to test your Flow:

```
class MovieViewModelTest {
    ...

    @Test
    fun fetchMoviesError() {
        ...
        val exception = "Test Exception"
        val movieRepository: MovieRepository = mock {
            onBlocking { fetchMovies() } doAnswer
                flow {
                    throw RuntimeException(exception)
                }
        }//mock
        val movieViewModel =
```

```
                MovieViewModel(movieRepository)


        runTest {
            movieViewModel.fetchMovies().test {
                assertEquals(exception,
                    awaitError().message)
            }
        }
    }
}
```

In this example, we used the `awaitError()` function to receive the exception and compare its message with the expected exception.

To test hot flows, you have to emit values inside the `test` lambda. You can also use the `cancelAndConsumeRemainingEvents()` function or the `cancelAndIgnoreRemainingEvents()` function to cancel any remaining events from the Flow.

The following shows an example of using the `cancelAndIgnoreRemainingEvents()` function after checking the first item from the Flow:

```
class MovieViewModelTest {
    ...

    @Test
    fun fetchMovies() {
        ...
        val expectedList = listOf(movie1, movie2)

        val movieRepository: MovieRepository = mock {
            onBlocking { fetchMovies() } doReturn
                flowOf(expectedList)
        }
        val movieViewModel =
            MovieViewModel(movieRepository)


        runTest {
```

```
            movieViewModel.fetchMovies().test {
                assertEquals(movie1, awaitItem())
                cancelAndIgnoreRemainingEvents()
            }
        }
    }
}
```

Here, the test will check the first item from the Flow, ignore any remaining items, and cancel the Flow.

In this section, you have learned how to test Flows with Turbine. Let's try what we have learned so far by adding some tests to Flows in an Android project.

Exercise 7.01 – Adding tests to Flows in an Android app

For this exercise, you will be continuing the movie app you worked on in *Exercise 6.01 – Handling Flow exception in an Android app*. This application displays the movies that are currently playing in movie theatres. You will be adding tests for the Kotlin Flows in the project by following these steps:

1. Open in Android Studio the movie app you worked on in *Exercise 6.01 – Handling Flow exception in an Android app*.

2. Go to the `MovieViewModelTest` class. Run the test class, and the `fetchMovies()` test function will fail. That is because we changed the implementation to use Flow in the previous chapter.

3. Remove the content of the `fetchMovies()` test function and replace it with the following content:

```
@Test
fun fetchMovies() {
    val dispatcher = StandardTestDispatcher()

    val movies = listOf(Movie(title = "Movie"))
    val expectedMovies =
      MutableLiveData<List<Movie>>()
    expectedMovies.postValue(movies)

    val movieRepository: MovieRepository = mock {
        onBlocking { fetchMoviesFlow() } doReturn
          flowOf(movies)
```

```
    }

    val movieViewModel =
        MovieViewModel(movieRepository, dispatcher)
}
```

With this code, we will be mocking `MovieRepository` to return a Flow of a list of movies, `movies`, which contains a single movie.

4. At the end of the `fetchMovies()` function, add the following code to test the `fetchMovies()` function of `MovieViewModel`:

```
@Test
fun fetchMovies() {

    ...


    runTest {
        movieViewModel.fetchMovies()
        dispatcher.scheduler.advanceUntilIdle()
        assertEquals(expectedMovies.value,
          movieViewModel.movies.value)

    }
}
```

This will call the `fetchMovies()` function from `movieViewModel`. We will then compare the returned `movieViewModel.movies` to see whether they are the same as the expected `movies` list (with a single Movie item).

5. In the `loading()` test function, replace the assertions with the following:

```
assertTrue(movieViewModel.loading.value)
dispatcher.scheduler.advanceUntilIdle()
assertFalse(movieViewModel.loading.value)
```

The `loading` variable is no longer nullable, so this simplifies the assertion statements.

6. Run the `MovieViewModelTest` class again. It should successfully run, and all the tests will pass.

7. Open the `MovieRepositoryTest` class. We will be adding tests for the
    `fetchMoviesFlow()` function of `MovieRepository`. First, add the following
    function to test the successful case of the function:

```
@Test
fun fetchMoviesFlow() {
    val movies = listOf(Movie(id = 3), Movie(id = 4))
    val response = MoviesResponse(1, movies)

    val movieService: MovieService = mock {
        onBlocking { getMovies(anyString()) } doReturn
            response
    }
    val movieRepository =
      MovieRepository(movieService)

    runTest {
        movieRepository.fetchMoviesFlow().collect {
            assertEquals(movies, it)
        }
    }
}
```

This will mock `MovieRepository` to always return the list of movies that we will later
compare with the movies from the `fetchMoviesFlow()` function.

8. Add the following function to add a test for the case when the `fetchMoviesFlow()`
    function throws an exception:

```
@Test
fun fetchMoviesFlowWithError() {
    val exception = "Test Exception"

    val movieService: MovieService = mock {
        onBlocking { getMovies(anyString()) } doThrow
            RuntimeException(exception)
    }
```

```
        val movieRepository =
          MovieRepository(movieService)

        runTest {
            movieRepository.fetchMoviesFlow().catch {
                assertEquals(exception, it.message)
            }
        }
    }
```

This test will use a fake `MovieRepository` that will always throw an error when calling `fetchMoviesFlow`. We will then test whether the exception thrown will be the same as the one that we expect.

9. Run the `MovieRepositoryTest` class. All the tests in `MovieRepository Test` should run and pass without an error.

10. Now, we will use the Turbine testing library to test the Flow from the `fetchMoviesFlow()` function of `MovieRepository`. Add the following in the `app/build.gradle` dependencies:

```
testImplementation 'app.cash.turbine:turbine:0.8.0'
```

This will allow us to use the Turbine testing library to create unit tests for Flows in our Android project.

11. Add a new test function to test the success case of the `fetchMoviesFlow()` function by adding the following:

```
@Test
fun fetchMoviesFlowTurbine() {
    val movies = listOf(Movie(id = 3), Movie(id = 4))
    val response = MoviesResponse(1, movies)

    val movieService: MovieService = mock {
        onBlocking { getMovies(anyString()) } doReturn
            response
    }
    val movieRepository =
      MovieRepository(movieService)

    runTest {
```

```
            movieRepository.fetchMoviesFlow().test {
                assertEquals(movies, awaitItem())
                awaitComplete()
            }
        }
    }
```

With this, we will be mocking MovieRepository to return a list of movies. We will later compare that with the list from movieRepository.fetchMoviesFlow() using awaitItem(). The awaitComplete() function will then check that the Flow has terminated.

12. Add another function to test using Turbine in the case when fetchMoviesFlow throws an exception by adding the following:

```
@Test
fun fetchMoviesFlowWithErrorTurbine() {
    val exception = "Test Exception"

    val movieService: MovieService = mock {
        onBlocking { getMovies(anyString()) } doThrow
            RuntimeException(exception)
    }
    val movieRepository =
        MovieRepository(movieService)

    runTest {
        movieRepository.fetchMoviesFlow().test {
            assertEquals(exception,
                awaitError().message)
        }
    }
}
```

This will use a MovieRepository mock class that will throw RuntimeException when calling fetchMoviesFlow(). We will then test that the exception message is the same one that was fetched, using the awaitError() call.

13. Run the MovieRepositoryTest class again. All the tests in MovieRepository Test should run and pass without an error.

In this exercise, we have worked on an Android project that uses Kotlin Flow, and we have created tests for these Flows.

Summary

This chapter focused on testing Kotlin Flows in our Android project. We started by setting up the project for adding tests for the Flows. The coroutines testing library (**kotlinx-coroutines-test**) can help you in creating tests for coroutines and Flows.

We learned how to add tests for the Flows in your Android application. You can use a mock class that returns a Flow of values and then compare it with the returned values. You can also convert a Flow into `List` or `Set`, or take values from the Flow; you can then compare them with the expected values.

Then, we learned about testing hot Flows with Turbine, a third-party testing library for testing Kotlin Flows. Turbine has a `test` extension on Flow where you can consume and compare values one by one.

Finally, we worked on an exercise where we created tests for the Kotlin Flows in an existing Android project. We also used the Turbine testing library to make the writing of tests for Flows easier.

Throughout the book, we have gained knowledge and skills about asynchronous programming in Android. We learned how to use Kotlin coroutines and Flow to simplify asynchronous programming in our Android projects.

Everything in Android is always evolving. There are also more advanced topics about coroutines and Flow that we have not covered. It is good to keep yourself up to date with the latest updates about Android, Kotlin coroutines, and Kotlin Flow. You can find out the latest about coroutines on Android at `https://developer.android.com/kotlin/coroutines` and the latest about Kotlin Flow on Android at `https://developer.android.com/kotlin/flow`.

Index

`Packt.com`

Subscribe to our online digital library for full access to over 7,000 books and videos, as well as industry leading tools to help you plan your personal development and advance your career. For more information, please visit our website.

Why subscribe?

- Spend less time learning and more time coding with practical eBooks and Videos from over 4,000 industry professionals

- Improve your learning with Skill Plans built especially for you

- Get a free eBook or video every month

- Fully searchable for easy access to vital information

- Copy and paste, print, and bookmark content

Did you know that Packt offers eBook versions of every book published, with PDF and ePub files available? You can upgrade to the eBook version at `packt.com` and as a print book customer, you are entitled to a discount on the eBook copy. Get in touch with us at `customercare@packtpub.com` for more details.

At `www.packt.com`, you can also read a collection of free technical articles, sign up for a range of free newsletters, and receive exclusive discounts and offers on Packt books and eBooks.

Other Books You May Enjoy

If you enjoyed this book, you may be interested in these other books by Packt:

Clean Android Architecture

Alexandru Dumbravan

ISBN: 978-1-80323-458-8

- Discover and solve issues in Android legacy applications
- Become well versed in the principles behind clean architecture
- Get to grips with writing loosely coupled and testable code
- Find out how to structure an application's code in separate layers
- Understand the role each layer plays in keeping the application clean

Kickstart Modern Android Development with Jetpack and Kotlin

Catalin Ghita

ISBN: 978-1-80181-107-1

- Integrate popular Jetpack libraries such as Compose, ViewModel, Hilt, and Navigation into real Android apps with Kotlin
- Apply modern app architecture concepts such as MVVM, dependency injection, and clean architecture
- Explore Android libraries such as Retrofit, Coroutines, and Flow
- Integrate Compose with the rest of the Jetpack libraries or other popular Android libraries
- Work with other Jetpack libraries such as Paging and Room while integrating a real REST API that supports pagination

Packt is searching for authors like you

If you're interested in becoming an author for Packt, please visit `authors.packtpub.com` and apply today. We have worked with thousands of developers and tech professionals, just like you, to help them share their insight with the global tech community. You can make a general application, apply for a specific hot topic that we are recruiting an author for, or submit your own idea.

Share Your Thoughts

Hi!

I am Jomar Tigcal, author of *Simplifying Android Development with Coroutines and Flows*. I really hope you enjoyed reading this book and found it useful for increasing your productivity and efficiency in Coroutines and Flows.

It would really help me (and other potential readers!) if you could leave a review on Amazon sharing your thoughts on *Simplifying Android Development with Coroutines and Flows*.

Go to the link below or scan the QR code to leave your review:

`https://packt.link/r/1801816247`

Your review will help me to understand what's worked well in this book, and what could be improved upon for future editions, so it really is appreciated.

Best Wishes,

Jomar Tigcal

www.ingramcontent.com/pod-product-compliance
Lightning Source LLC
Chambersburg PA
CBHW060141060326
40690CB00018B/3942